James J. Buckley

Simulating Fuzzy Systems

Studies in Fuzziness and Soft Computing, Volume 171

Editor-in-chief
Prof. Janusz Kacprzyk
Systems Research Institute
Polish Academy of Sciences
ul. Newelska 6
01-447 Warsaw
Poland
E-mail: kacprzyk@ibspan.waw.pl

James J. Buckley

Simulating Fuzzy Systems

Springer

Professor James J. Buckley
University of Alabama at Birmingham
Department of Mathematics
Birmingham, AL 35294-1170
USA
E-mail: buckley@math.uab.edu

ISSN print edition: 1434-9922
ISSN electronic edition: 1860-0808
ISBN 3-642-42534-8 Springer Berlin Heidelberg New York

Springer is a part of Springer Science+Business Media
springeronline.com

© Springer-Verlag Berlin Heidelberg 2005
Softcover re-print of the Hardcover 1st edition 2005

Typesetting: by the author and TechBooks using a Springer LaTeX macro package
Cover design: E. Kirchner, Springer Heidelberg
Printed on acid-free paper   62/3141/jl- 5 4 3 2 1 0

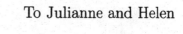

To Julianne and Helen

Contents

1 Introduction

1.1 Introduction

This book is written in two major parts. The first part includes the introductory chapters consisting of Chaps. 1 through 8. In part two, Chaps. 9–26, we present the applications.

First we need to be familiar with fuzzy sets. All you need to know about fuzzy sets for this book comprises Chap. 2. For a beginning introduction to fuzzy sets and fuzzy logic see [6].

Chapter 3 gives a brief introduction to fuzzy estimation. We explain how you can get fuzzy numbers when you estimate, from crisp data, probabilities or parameters in probability densities. The basic construction involves placing confidence intervals, one on top of another, to obtain a fuzzy number as our estimator instead of using a point estimator or a single confidence interval.

Fuzzy probabilities and fuzzy parameters in probability densities gives rise to fuzzy distributions. In Chap. 4 we look more closely at the fuzzy binomial, fuzzy Poisson, fuzzy normal, fuzzy exponential and the fuzzy uniform distribution. These are the fuzzy distributions that we will be using in the rest of the book. For all of these fuzzy distributions we want to see how they are used to compute fuzzy probabilities.

Chapter 5 introduces fuzzy systems theory. Consider any system whose description employs probability theory. Assume that some of these probabilities must be estimated from crisp data, and/or some the parameters in the probability densities must be estimated from crisp data. Using our fuzzy estimators (Chap. 3) we obtain fuzzy probabilities and fuzzy distributions. Now we want to compute certain system descriptors like $R =$ the expected time it takes an item to pass through the system and $N =$ the expected number of items in the system. The fuzziness in the probabilities and probability densities propagates through the system producing fuzzy numbers for R and N. We have a fuzzy system. A main problem with fuzzy systems [2,9] is: how will we accomplish all the needed fuzzy calculations to compute the system descriptors? This is the main topic of this book: use crisp simulation to estimate the fuzzy numbers describing system performance.

How do we choose simulation software to accomplish all the simulations in Chaps. 7, 9–26 is the topic of Chap. 6. We discuss cost, ease of use, need to

James J. Buckley: *Simulating Fuzzy Systems*, StudFuzz **171**, 1–4 (2005)
www.springerlink.com

run on a desktop computer, must compile system statistics we are interested in, plus some others. Our final decision is also discussed.

Simulation can not be used to approximate all fuzzy calculations. Which fuzzy computations can be approximated is the topic of Chap. 7. Consider a very simple queuing system where customers arrive and possibly enter a queue waiting for service in a single server and after service they depart the system. If the server is vacant the customer can go directly into the server. We want to find N = the expected number of customers in the system. The input to the model is λ = the arrival rate and μ = the service rate. In any operations research book we may find a function f to compute N in terms of λ and μ, or $N = f(\lambda, \mu)$. Assume that λ and μ must be estimated from data on the system. We obtain fuzzy estimators which incorporate the uncertainty in the data. So now λ and μ become fuzzy numbers. Put the fuzzy numbers into the function f and determine, using the extension principle (Chap. 2), the fuzzy number for N. We argue in Chap. 7 that simulation can estimate alpha-cuts (horizontal slices, Chap. 2) of N. We say that N can be gotten in a one-step calculation since we have N a function of λ and μ. Next consider finding N using a two-step calculation. Here we first find the steady-state probabilities from the λ and μ and then get N as a function of the steady-state probabilities. Then λ and μ become fuzzy, the steady-state probabilities are fuzzy and finally we get fuzzy N. In this case N is obtained as the result of a two-step calculation. We argue in Chap. 7 that simulation may not approximate horizontal cuts of fuzzy N from a two-step (or more) calculation.

Chapter 8 introduces a type of simulation optimization. We discuss how we plan to solve the simulation optimization problems presented in (5.4) and (5.5) of Chap. 5, and (7.20) and (7.21) of Chap. 7. This solution must be performed in all Chaps. 9–26.

The structure of the rest of the book is now determined. Use simulation to approximate fuzzy system descriptors in fuzzy systems where the fuzzy number to be approximated can be assumed to be the result of a one-step calculation but the system is sufficiently complicated so that we do not know this one-step function. This will be the topic of Chaps. 9–26.

The applications in Chaps. 9–26 are quite varied ranging from emergency rooms to machine shops to project scheduling showing the varieties of fuzzy systems. These chapters may be read independently (except Chaps. 11,12 and 16,17 go together). This means some material, including discussion of steady-state, fuzzy estimators, one-step functions, etc., are repeated in each chapter.

Selected simulation programs are in Chap. 28. These simulation programs, including minimal comments, are listed in that chapter. We could not include all the programs because that would make the Chap. 40 pages long, too much devoted to computer programs in a book around 200 pages. A reader may obtain other simulation programs by contacting the author via email with their requests.

This book is based on, but expanded from, the following recent papers and publications: (1) fuzzy estimation, probability and statistics [1–4], [7–9]; (2) fuzzy systems [5]; and (3) simulating fuzzy systems [10, 11].

There are no prerequisites, but it would be helpful to know some basic information about queuing systems. However, the reader should be able to understand, from the figures and analytical development, how simulation is useful in analyzing fuzzy systems.

1.2 Notation

It is difficult, in a book with a lot of mathematics, to achieve a uniform notation without having to introduce many new specialized symbols. Our basic notation is presented in Chap. 2. What we have done is to have a uniform notation within each chapter. What this means is that we may use the letters "a" and "b" to represent a closed interval $[a, b]$ in one chapter but they could stand for parameters in a probability density in another chapter. We will have the following uniform notation throughout the book: (1) we place a "bar" over a letter to denote a fuzzy set ($\overline{A}$, $\overline{B}$, etc.), and all our fuzzy sets will be fuzzy subsets of the real numbers; and (2) an alpha-cut of a fuzzy set (Chap. 2) is always denoted by "α". Since we will be using α for alpha-cuts we need to change some standard notation in statistics: we use β in confidence intervals. So a $(1 - \beta)100\%$ confidence interval means a 95% confidence interval if $\beta = 0.05$. When a confidence interval switches to being an alpha-cut of a fuzzy number (see Chap. 3), we switch from β to α. All fuzzy arithmetic is performed using the extension principle (Chap. 2). The term "crisp" means not fuzzy. A crisp set is a regular set and a crisp number is a real number. Also, throughout the book $\overline{x}$ will be the mean of a random sample, not a fuzzy set.

1.3 Figures

Some of the figures, graphs of certain fuzzy numbers, in the book are difficult to obtain so they were created using different methods. Many graphs were done first in Maple [12] and then exported to $LaTeX2_\epsilon$. We did these figures first in Maple because of the "implicitplot" command in Maple. Let us explain why this command was important in this book. Suppose $\overline{X}$ is a fuzzy estimator we want to graph. Usually in this book we determine $\overline{X}$ by first calculating its α-cuts. Let $\overline{X}[\alpha] = [x_1(\alpha), x_2(\alpha)]$. So we get $x = x_1(\alpha)$ describing the left side of the triangular shaped fuzzy number $\overline{X}$ and $x = x_2(\alpha)$ describes the right side. On a graph we would have the x-axis horizontal and the y-axis vertical. α is on the y-axis between zero and one. Substituting y

for α we need to graph $x = x_i(y)$, for $i = 1, 2$. But this is backwards, we usually have y a function of x. The "implicitplot" command allows us to do the correct graph with x a function of y when we have $x = x_i(y)$. Figures 2.1–2.3, 3.1–3.7, 4.1, 4.3, and 4.5 were done in Maple and then exported to $LaTeX2_\epsilon$. All the other figures were constructed in $LaTeX2_\epsilon$.

References

1. J.J. Buckley: Fuzzy Probabilities: New Approach and Applications, Physica-Verlag, Heidelberg, Germany, 2003.
2. J.J. Buckley: Fuzzy Probabilities and Fuzzy Sets for Web Planning, Springer, Heidelberg, Germany, 2004.
3. J.J. Buckley: Fuzzy Statistics, Springer, Heidelberg, Germany, 2004.
4. J.J. Buckley: Uncertain Probabilities III: The Continuous Case, Soft Computing, 8(2004)200-206.
5. J.J. Buckley: Fuzzy Systems, Soft Computing. To appear.
6. J.J. Buckley and E. Eslami: An Introduction to Fuzzy Logic and Fuzzy Sets, Physica-Verlag, Heidelberg, Germany, 2002.
7. J.J. Buckley and E. Eslami: Uncertain Probabilities I: The Discrete Case, Soft Computing, 7(2003)500-505.
8. J.J. Buckley and E. Eslami: Uncertain Probabilities II: The Continuous Case, Soft Computing, 8(2004)193-199.
9. J.J. Buckley, K. Reilly and X. Zheng: Fuzzy Probabilities for Web Planning, Soft Computing, 8(2004)464-476.
10. J.J. Buckley, K. Reilly and X. Zheng: Simulating Fuzzy Systems I, in: Applied Research in Uncertainty Modelling and Analysis, Eds. N.O. Attoh-Okine, B. Ayyub, Kluwer, 2004. To appear.
11. J.J. Buckley, K. Reilly and X. Zheng: Simulating Fuzzy Systems II, in: Applied Research in Uncertainty Modelling and Analysis, Eds. N.O. Attoh-Okine, B. Ayyub, Kluwer, 2004. To appear.
12. Maple 9, Waterloo Maple Inc., Waterloo, Canada.

2 Fuzzy Sets

2.1 Introduction

In this chapter we have collected together the basic ideas from fuzzy sets and fuzzy functions needed for the book. Any reader familiar with fuzzy sets, fuzzy numbers, the extension principle, α-cuts, interval arithmetic, and fuzzy functions may go on and have a look at Sects. 2.5, 2.6 and 2.7. In Sect. 2.5 we discuss the method we will be using in this book to evaluate comparisons between fuzzy numbers. That is, in Sect. 2.5 we need to decide which one of the following three possibilities is true: $\overline{M} < \overline{N}$; $\overline{M} \approx \overline{N}$; or $\overline{M} > \overline{N}$, for fuzzy numbers $\overline{M}$ and $\overline{N}$. In Sect. 2.6, using Sect. 2.5, we solve, for small sets of fuzzy numbers, $max\overline{M}_i$ and $min\overline{M}_i$, $1 \leq i \leq n$. Section 2.6 explains how, and why, we will approximate discrete fuzzy sets with fuzzy numbers. A good general reference for fuzzy sets and fuzzy logic is [4] and [8].

Our notation specifying a fuzzy set is to place a "bar" over a letter. So $\overline{X}, \overline{M}, \overline{T}, \ldots, \overline{\mu}, \overline{p}, \overline{\sigma}^2, \overline{a}, \overline{b}, \ldots$, all denote fuzzy sets.

2.2 Fuzzy Sets

If Ω is some set, then a fuzzy subset $\overline{A}$ of Ω is defined by its membership function, written $\overline{A}(x)$, which produces values in $[0,1]$ for all x in Ω. So, $\overline{A}(x)$ is a function mapping Ω into $[0,1]$. If $\overline{A}(x_0) = 1$, then we say x_0 belongs to $\overline{A}$, if $\overline{A}(x_1) = 0$ we say x_1 does not belong to $\overline{A}$, and if $\overline{A}(x_2) = 0.6$ we say the membership value of x_2 in $\overline{A}$ is 0.6. When $\overline{A}(x)$ is always equal to one or zero we obtain a crisp (non-fuzzy) subset of Ω. For all fuzzy sets $\overline{B}, \overline{C}, \ldots$ we use $\overline{B}(x), \overline{C}(x), \ldots$ for the value of their membership function at x. The fuzzy sets we will be using will be fuzzy numbers .

The term "crisp" will mean not fuzzy. A crisp set is a regular set. A crisp number is just a real number. A crisp function maps real numbers (or real vectors) into real numbers. A crisp solution to a problem is a solution involving crisp sets, crisp numbers, crisp functions, etc.

2.2.1 Fuzzy Numbers

A general definition of fuzzy number may be found in [4, 8], however our fuzzy numbers will be triangular (shaped) fuzzy numbers. A triangular fuzzy

James J. Buckley: *Simulating Fuzzy Systems*, StudFuzz **171**, 5–18 (2005)
www.springerlink.com

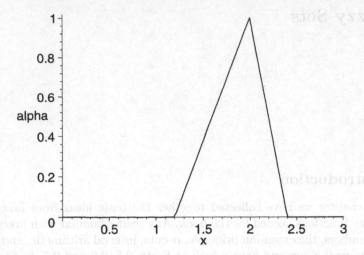

Fig. 2.1. Triangular Fuzzy Number $\overline{N}$

number $\overline{N}$ is defined by three numbers $a < b < c$ where the base of the triangle is the interval $[a, c]$ and its vertex is at $x = b$. Triangular fuzzy numbers will be written as $\overline{N} = (a/b/c)$. A triangular fuzzy number $\overline{N} = (1.2/2/2.4)$ is shown in Fig. 2.1. We see that $\overline{N}(2) = 1$, $\overline{N}(1.6) = 0.5$, etc.

A triangular shaped fuzzy number $\overline{P}$ is given in Fig. 2.2. $\overline{P}$ is only partially specified by the three numbers 1.2, 2, 2.4 since the graph on $[1.2, 2]$, and $[2, 2.4]$, is not a straight line segment. To be a triangular shaped fuzzy number we require the graph to be continuous and: (1) monotonically increasing on

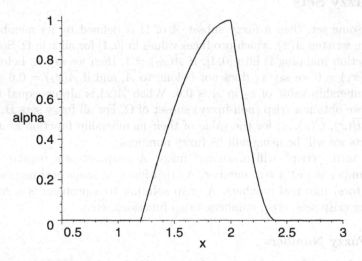

Fig. 2.2. Triangular Shaped Fuzzy Number $\overline{P}$

[1.2, 2]; and (2) monotonically decreasing on [2, 2.4]. For triangular shaped fuzzy number $\overline{P}$ we use the notation $\overline{P} \approx (1.2/2/2.4)$ to show that it is partially defined by the three numbers 1.2, 2, and 2.4. If $\overline{P} \approx (1.2/2/2.4)$ we know its base is on the interval [1.2, 2.4] with vertex (membership value one) at $x = 2$.

2.2.2 Alpha-Cuts

Alpha-cuts are slices through a fuzzy set producing regular (non-fuzzy) sets. If $\overline{A}$ is a fuzzy subset of some set Ω, then an α-cut of $\overline{A}$, written $\overline{A}[\alpha]$, is defined as

$$\overline{A}[\alpha] = \{x \in \Omega | \overline{A}(x) \geq \alpha\}, \tag{2.1}$$

for all α, $0 < \alpha \leq 1$. The $\alpha = 0$ cut, or $\overline{A}[0]$, must be defined separately.

Let $\overline{N}$ be the fuzzy number in Fig. 2.1. Then $\overline{N}[0] = [1.2, 2.4]$. Notice that using (2.1) to define $\overline{N}[0]$ would give $\overline{N}[0] =$ all the real numbers. Similarly, in Fig. 2.2 $\overline{P}[0] = [1.2, 2.4]$. For any fuzzy set $\overline{A}$, $\overline{A}[0]$ is called the support, or base, of $\overline{A}$. Many authors call the support of a fuzzy number the open interval (a, b) like the support of $\overline{N}$ in Fig. 2.1 would then be $(1.2, 2.4)$. However in this book we use the closed interval $[a, b]$ for the support (base) of the fuzzy number.

The core of a fuzzy number is the set of values where the membership value equals one. If $\overline{N} = (a/b/c)$, or $\overline{N} \approx (a/b/c)$, then the core of $\overline{N}$ is the single point b.

For any fuzzy number $\overline{Q}$ we know that $\overline{Q}[\alpha]$ is a closed, bounded, interval for $0 \leq \alpha \leq 1$. We will write this as

$$\overline{Q}[\alpha] = [q_1(\alpha), q_2(\alpha)], \tag{2.2}$$

where $q_1(\alpha)$ $(q_2(\alpha))$ will be an increasing (decreasing) function of α with $q_1(1) = q_2(1)$. If $\overline{Q}$ is a triangular shaped then: (1) $q_1(\alpha)$ will be a continuous, monotonically increasing function of α in $[0, 1]$; (2) $q_2(\alpha)$ will be a continuous, monotonically decreasing function of α, $0 \leq \alpha \leq 1$; and (3) $q_1(1) = q_2(1)$.

For the $\overline{N}$ in Fig. 2.1 we obtain $\overline{N}[\alpha] = [n_1(\alpha), n_2(\alpha)]$, $n_1(\alpha) = 1.2 + 0.8\alpha$ and $n_2(\alpha) = 2.4 - 0.4\alpha$, $0 \leq \alpha \leq 1$. The equation for $n_i(\alpha)$ is backwards. With the y-axis vertical and the x-axis horizontal the equation $n_1(\alpha) = 1.2 + 0.8\alpha$ means $x = 1.2 + 0.8y$, $0 \leq y \leq 1$. That is, the straight line segment from $(1.2, 0)$ to $(2, 1)$ in Fig. 2.1 is given as x a function of y whereas it is usually stated as y a function of x. This is how it will be done for all α-cuts of fuzzy numbers.

The general requirements for a fuzzy set $\overline{N}$ of the real numbers to be a fuzzy number are: (1) it must be normalized, or $\overline{N}(x) = 1$ for some x; and (2) its alpha-cuts must be closed, bounded, intervals for all alpha in $[0, 1]$. This will be important in fuzzy estimation because there the fuzzy numbers will have very short vertical line segments at both ends of its base (see Sect. 3.3 in Chap. 3). Even so, such a fuzzy set still meets the general requirements presented above to be called a fuzzy number.

2.2.3 Inequalities

Let $\overline{N} = (a/b/c)$. We write $\overline{N} \geq \delta$, δ some real number, if $a \geq \delta$, $\overline{N} > \delta$ when $a > \delta$, $\overline{N} \leq \delta$ for $c \leq \delta$ and $\overline{N} < \delta$ if $c < \delta$. We use the same notation for triangular shaped fuzzy numbers whose support is the interval $[a, c]$.

If $\overline{A}$ and $\overline{B}$ are two fuzzy subsets of a set Ω, then $\overline{A} \leq \overline{B}$ means $\overline{A}(x) \leq \overline{B}(x)$ for all x in Ω, or $\overline{A}$ is a fuzzy subset of $\overline{B}$. $\overline{A} < \overline{B}$ holds when $\overline{A}(x) < \overline{B}(x)$, for all x. There is a potential problem with the symbol $<$. In some places in the book, for example see Sect. 2.5, $\overline{M} < \overline{N}$ for fuzzy numbers $\overline{M}$ and $\overline{N}$ means that $\overline{M}$ is less than $\overline{N}$. It should be clear on how we use "$<$" as to which meaning is correct.

2.2.4 Discrete Fuzzy Sets

Let $\overline{A}$ be a fuzzy subset of Ω. If $\overline{A}(x)$ is not zero only at a finite number of x values in Ω, then $\overline{A}$ is called a discrete fuzzy set. Suppose $\overline{A}(x)$ is not zero only at x_1, x_2, x_3 and x_4 in Ω. Then we write the fuzzy set as

$$\overline{A} = \left\{ \frac{\mu_1}{x_1}, \ldots, \frac{\mu_4}{x_4} \right\}, \qquad (2.3)$$

where the μ_i are the membership values. That is, $\overline{A}(x_i) = \mu_i$, $1 \leq i \leq 4$, and $\overline{A}(x) = 0$ otherwise. We can have discrete fuzzy subsets of any space Ω. Notice that α-cuts of discrete fuzzy sets of $\mathbb{R}$, the set of real numbers, do not produce closed, bounded, intervals.

2.3 Fuzzy Arithmetic

If $\overline{A}$ and $\overline{B}$ are two fuzzy numbers we will need to add, subtract, multiply and divide them. There are two basic methods of computing $\overline{A} + \overline{B}$, $\overline{A} - \overline{B}$, etc. which are: (1) extension principle; and (2) α-cuts and interval arithmetic.

2.3.1 Extension Principle

Let $\overline{A}$ and $\overline{B}$ be two fuzzy numbers. If $\overline{A} + \overline{B} = \overline{C}$, then the membership function for $\overline{C}$ is defined as

$$\overline{C}(z) = \sup_{x,y}\{\min(\overline{A}(x), \overline{B}(y))|x + y = z\}. \qquad (2.4)$$

If we set $\overline{C} = \overline{A} - \overline{B}$, then

$$\overline{C}(z) = \sup_{x,y}\{\min(\overline{A}(x), \overline{B}(y))|x - y = z\}. \qquad (2.5)$$

Similarly, $\overline{C} = \overline{A} \cdot \overline{B}$, then

$$\overline{C}(z) = \sup_{x,y}\{\min(\overline{A}(x),\overline{B}(y))|x \cdot y = z\} \, , \tag{2.6}$$

and if $\overline{C} = \overline{A}/\overline{B}$,

$$\overline{C}(z) = \sup_{x,y}\{\min(\overline{A}(x),\overline{B}(y))|x/y = z\} \, . \tag{2.7}$$

In all cases $\overline{C}$ is also a fuzzy number [8]. We assume that zero does not belong to the support of $\overline{B}$ in $\overline{C} = \overline{A}/\overline{B}$. If $\overline{A}$ and $\overline{B}$ are triangular (shaped) fuzzy numbers then so are $\overline{A} + \overline{B}$ and $\overline{A} - \overline{B}$, but $\overline{A} \cdot \overline{B}$ and $\overline{A}/\overline{B}$ will be triangular shaped fuzzy numbers.

We should mention something about the operator "sup" in (2.3)–(2.6). If Ω is a set of real numbers bounded above (there is a M so that $x \leq M$, for all x in Ω), then $\sup(\Omega) =$ the least upper bound for Ω. If Ω has a maximum member, then $\sup(\Omega) = \max(\Omega)$. For example, if $\Omega = [0,1)$, $\sup(\Omega) = 1$ but if $\Omega = [0,1]$, then $\sup(\Omega) = \max(\Omega) = 1$. The dual operator to "sup" is "inf". If Ω is bounded below (there is a M so that $M \leq x$ for all $x \in \Omega$), then $\inf(\Omega) =$ the greatest lower bound. For example, for $\Omega = (0,1]$ $\inf(\Omega) = 0$ but if $\Omega = [0,1]$, then $\inf(\Omega) = \min(\Omega) = 0$.

Obviously, given $\overline{A}$ and $\overline{B}$, (2.3)–(2.6) appear quite complicated to compute $\overline{A} + \overline{B}$, $\overline{A} - \overline{B}$, etc. So, we now present another procedure based on α-cuts and interval arithmetic. First, we present the basics of interval arithmetic.

2.3.2 Interval Arithmetic

We only give a brief introduction to interval arithmetic. For more information the reader is referred to [9,10]. Let $[a_1, b_1]$ and $[a_2, b_2]$ be two closed, bounded, intervals of real numbers. If $*$ denotes addition, subtraction, multiplication, or division, then $[a_1, b_1] * [a_2, b_2] = [\alpha, \beta]$ where

$$[\alpha, \beta] = \{a * b | a_1 \leq a \leq b_1, a_2 \leq b \leq b_2\} \, . \tag{2.8}$$

If $*$ is division, we must assume that zero does not belong to $[a_2, b_2]$. We may simplify (2.8) as follows:

$$[a_1, b_1] + [a_2, b_2] = [a_1 + a_2, b_1 + b_2] \, , \tag{2.9}$$

$$[a_1, b_1] - [a_2, b_2] = [a_1 - b_2, b_1 - a_2] \, , \tag{2.10}$$

$$[a_1, b_1] / [a_2, b_2] = [a_1, b_1] \cdot \left[\frac{1}{b_2}, \frac{1}{a_2}\right] \, , \tag{2.11}$$

and

$$[a_1, b_1] \cdot [a_2, b_2] = [\alpha, \beta] \, , \tag{2.12}$$

where

$$\alpha = \min\{a_1 a_2, a_1 b_2, b_1 a_2, b_1 b_2\} \, , \tag{2.13}$$

$$\beta = \max\{a_1 a_2, a_1 b_2, b_1 a_2, b_1 b_2\} \, . \tag{2.14}$$

Multiplication and division may be further simplified if we know that $a_1 > 0$ and $b_2 < 0$, or $b_1 > 0$ and $b_2 < 0$, etc. For example, if $a_1 \geq 0$ and $a_2 \geq 0$, then

$$[a_1, b_1] \cdot [a_2, b_2] = [a_1 a_2, b_1 b_2] , \tag{2.15}$$

and if $b_1 < 0$ but $a_2 \geq 0$, we see that

$$[a_1, b_1] \cdot [a_2, b_2] = [a_1 b_2, a_2 b_1] . \tag{2.16}$$

Also, assuming $b_1 < 0$ and $b_2 < 0$ we get

$$[a_1, b_1] \cdot [a_2, b_2] = [b_1 b_2, a_1 a_2] , \tag{2.17}$$

but $a_1 \geq 0$, $b_2 < 0$ produces

$$[a_1, b_1] \cdot [a_2, b_2] = [a_2 b_1, b_2 a_1] . \tag{2.18}$$

2.3.3 Fuzzy Arithmetic

Again we have two fuzzy numbers $\overline{A}$ and $\overline{B}$. We know α-cuts are closed, bounded, intervals so let $\overline{A}[\alpha] = [a_1(\alpha), a_2(\alpha)]$, $\overline{B}[\alpha] = [b_1(\alpha), b_2(\alpha)]$. Then if $\overline{C} = \overline{A} + \overline{B}$ we have

$$\overline{C}[\alpha] = \overline{A}[\alpha] + \overline{B}[\alpha] . \tag{2.19}$$

We add the intervals using (2.8). Setting $\overline{C} = \overline{A} - \overline{B}$ we get

$$\overline{C}[\alpha] = \overline{A}[\alpha] - \overline{B}[\alpha] , \tag{2.20}$$

for all α in $[0, 1]$. Also

$$\overline{C}[\alpha] = \overline{A}[\alpha] \cdot \overline{B}[\alpha] , \tag{2.21}$$

for $\overline{C} = \overline{A} \cdot \overline{B}$ and

$$\overline{C}[\alpha] = \overline{A}[\alpha]/\overline{B}[\alpha] , \tag{2.22}$$

when $\overline{C} = \overline{A}/\overline{B}$, provided that zero does not belong to $\overline{B}[\alpha]$ for all α. This method is equivalent to the extension principle method of fuzzy arithmetic [8]. Obviously, this procedure, of α-cuts plus interval arithmetic, is more user (and computer) friendly.

Example 2.3.3.1

Let $\overline{A} = (-3/-2/-1)$ and $\overline{B} = (4/5/6)$. We determine $\overline{A} \cdot \overline{B}$ using α-cuts and interval arithmetic. We compute $\overline{A}[\alpha] = [-3 + \alpha, -1 - \alpha]$ and $\overline{B}[\alpha] = [4 + \alpha, 6 - \alpha]$. So, if $\overline{C} = \overline{A} \cdot \overline{B}$ we obtain $\overline{C}[\alpha] = [(\alpha - 3)(6 - \alpha), (-1 - \alpha)(4 + \alpha)]$, $0 \leq \alpha \leq 1$. The graph of $\overline{C}$ is shown in Fig. 2.3.

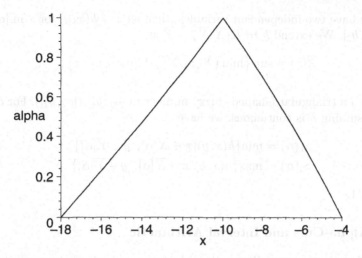

Fig. 2.3. The Fuzzy Number $\overline{C} = \overline{A} \cdot \overline{B}$

2.4 Fuzzy Functions

In this book a fuzzy function is a mapping from fuzzy numbers into fuzzy numbers. We write $H(\overline{X}) = \overline{Z}$ for a fuzzy function with one independent variable $\overline{X}$. $\overline{X}$ will be a triangular (shaped) fuzzy number and then we usually obtain $\overline{Z}$ as a triangular (shaped) shaped fuzzy number. For two independent variables we have $H(\overline{X}, \overline{Y}) = \overline{Z}$.

Where do these fuzzy functions come from? They are usually extensions of real-valued functions. Let $h : [a, b] \rightarrow \mathbb{R}$. This notation means $z = h(x)$ for x in $[a, b]$ and z a real number. One extends $h : [a, b] \rightarrow \mathbb{R}$ to $H(\overline{X}) = \overline{Z}$ in two ways: (1) the extension principle; or (2) using α-cuts and interval arithmetic.

2.4.1 Extension Principle

Any $h : [a, b] \rightarrow \mathbb{R}$ may be extended to $H(\overline{X}) = \overline{Z}$ as follows

$$\overline{Z}(z) = \sup_{x} \left\{ \overline{X}(x) | h(x) = z, \ a \le x \le b \right\} . \tag{2.23}$$

Equation (2.22) defines the membership function of $\overline{Z}$ for any triangular (shaped) fuzzy number $\overline{X}$ in $[a, b]$.

If h is continuous, then we have a way to find α-cuts of $\overline{Z}$. Let $\overline{Z}[\alpha] = [z_1(\alpha), z_2(\alpha)]$. Then [5]

$$z_1(\alpha) = \min\{h(x) | x \in \overline{X}[\alpha]\} , \tag{2.24}$$
$$z_2(\alpha) = \max\{h(x) | x \in \overline{X}[\alpha]\} , \tag{2.25}$$

for $0 \le \alpha \le 1$.

If we have two independent variables, then let $z = h(x, y)$ for x in $[a_1, b_1]$, y in $[a_2, b_2]$. We extend h to $H(\overline{X}, \overline{Y}) = \overline{Z}$ as

$$\overline{Z}(z) = \sup_{x,y} \left\{ \min \left(\overline{X}(x), \overline{Y}(y) \right) | h(x, y) = z \right\}, \tag{2.26}$$

for $\overline{X}$ ($\overline{Y}$) a triangular (shaped) fuzzy number in $[a_1, b_1]$ ($[a_2, b_2]$). For α-cuts of $\overline{Z}$, assuming h is continuous, we have

$$z_1(\alpha) = \min\{h(x, y) | x \in \overline{X}[\alpha], \ y \in \overline{Y}[\alpha]\}, \tag{2.27}$$
$$z_2(\alpha) = \max\{h(x, y) | x \in \overline{X}[\alpha], \ y \in \overline{Y}[\alpha]\}, \tag{2.28}$$

$0 \leq \alpha \leq 1$.

2.4.2 Alpha-Cuts and Interval Arithmetic

All the functions we usually use in engineering and science have a computer algorithm which, using a finite number of additions, subtractions, multiplications and divisions, can evaluate the function to required accuracy [5]. Such functions can be extended, using α-cuts and interval arithmetic, to fuzzy functions. Let $h : [a, b] \rightarrow \mathbb{R}$ be such a function. Then its extension $H(\overline{X}) = \overline{Z}$, $\overline{X}$ in $[a, b]$ is done, via interval arithmetic, in computing $h(\overline{X}[\alpha]) = \overline{Z}[\alpha]$, α in $[0, 1]$. We input the interval $\overline{X}[\alpha]$, perform the arithmetic operations needed to evaluate h on this interval, and obtain the interval $\overline{Z}[\alpha]$. Then put these α-cuts together to obtain the value $\overline{Z}$. The extension to more independent variables is straightforward.

For example, consider the fuzzy function

$$\overline{Z} = H(\overline{X}) = \frac{\overline{A}\,\overline{X} + \overline{B}}{\overline{C}\,\overline{X} + \overline{D}}, \tag{2.29}$$

for triangular fuzzy numbers $\overline{A}, \overline{B}, \overline{C}, \overline{D}$ and triangular fuzzy number $\overline{X}$ in $[0, 10]$. We assume that $\overline{C} \geq 0$, $\overline{D} > 0$ so that $\overline{C}\,\overline{X} + \overline{D} > 0$. This would be the extension of

$$h(x_1, x_2, x_3, x_4, x) = \frac{x_1 x + x_2}{x_3 x + x_4}. \tag{2.30}$$

We would substitute the intervals $\overline{A}[\alpha]$ for x_1, $\overline{B}[\alpha]$ for x_2, $\overline{C}[\alpha]$ for x_3, $\overline{D}[\alpha]$ for x_4 and $\overline{X}[\alpha]$ for x, do interval arithmetic, to obtain interval $\overline{Z}[\alpha]$ for $\overline{Z}$. Alternatively, the fuzzy function

$$\overline{Z} = H(\overline{X}) = \frac{2\overline{X} + 10}{3\overline{X} + 4}, \tag{2.31}$$

would be the extension of

$$h(x) = \frac{2x + 10}{3x + 4}. \tag{2.32}$$

2.4.3 Differences

Let $h : [a, b] \to \mathbb{R}$. Just for this subsection let us write $\overline{Z}^* = H(\overline{X})$ for the extension principle method of extending h to H for $\overline{X}$ in $[a, b]$. We denote $\overline{Z} = H(\overline{X})$ for the α-cut and interval arithmetic extension of h .

We know that $\overline{Z}$ can be different from $\overline{Z}^*$. But for basic fuzzy arithmetic in Sect. 2.3 the two methods give the same results. In the example below we show that for $h(x) = x(1 - x)$, x in $[0, 1]$, we can get $\overline{Z}^* \neq \overline{Z}$ for some $\overline{X}$ in $[0, 1]$. What is known ([5],[9]) is that for usual functions in science and engineering $\overline{Z}^* \leq \overline{Z}$. Otherwise, there is no known necessary and sufficient conditions on h so that $\overline{Z}^* = \overline{Z}$ for all $\overline{X}$ in $[a, b]$.

There is nothing wrong in using α-cuts and interval arithmetic to evaluate fuzzy functions. Surely, it is user, and computer friendly. However, we should be aware that whenever we use α-cuts plus interval arithmetic to compute $\overline{Z} = H(\overline{X})$ we may be getting something larger than that obtained from the extension principle. The same results hold for functions of two or more independent variables.

Example 2.4.3.1

The example is the simple fuzzy expression

$$\overline{Z} = (1 - \overline{X}) \, \overline{X} \,, \tag{2.33}$$

for $\overline{X}$ a triangular fuzzy number in $[0, 1]$. Let $\overline{X}[\alpha] = [x_1(\alpha), x_2(\alpha)]$. Using interval arithmetic we obtain

$$z_1(\alpha) = (1 - x_2(\alpha))x_1(\alpha) \,, \tag{2.34}$$
$$z_2(\alpha) = (1 - x_1(\alpha))x_2(\alpha) \,, \tag{2.35}$$

for $\overline{Z}[\alpha] = [z_1(\alpha), z_2(\alpha)]$, α in $[0, 1]$.

The extension principle extends the regular equation $z = (1 - x)x$, $0 \leq x \leq 1$, to fuzzy numbers as follows

$$\overline{Z}^*(z) = \sup_x \left\{ \overline{X}(x) | (1 - x)x = z, \ 0 \leq x \leq 1 \right\} \,. \tag{2.36}$$

Let $\overline{Z}^*[\alpha] = [z_1^*(\alpha), z_2^*(\alpha)]$. Then

$$z_1^*(\alpha) = \min\{(1 - x)x | x \in \overline{X}[\alpha]\} \,, \tag{2.37}$$
$$z_2^*(\alpha) = \max\{(1 - x)x | x \in \overline{X}[\alpha]\} \,, \tag{2.38}$$

for all $0 \leq \alpha \leq 1$. Now let $\overline{X} = (0/0.25/0.5)$, then $x_1(\alpha) = 0.25\alpha$ and $x_2(\alpha) = 0.50 - 0.25\alpha$. Equations (2.33) and (2.34) give $\overline{Z}[0.50] = [5/64, 21/64]$ but (2.36) and (2.37) produce $\overline{Z}^*[0.50] = [7/64, 15/64]$. Therefore, $\overline{Z}^* \neq \overline{Z}$. We do know that if each fuzzy number appears only once in the fuzzy expression, the two methods produce the same results [5, 9]. However, if a fuzzy number is used more than once, as in (2.32), the two procedures can give different results.

2.5 Ordering/Ranking Fuzzy Numbers

This section is about ordering/ranking a finite set of fuzzy numbers. Given a finite set of fuzzy numbers $\overline{A}_1, \ldots, \overline{A}_n$, we want to order/rank them from smallest to largest. For a finite set of real numbers there is no problem in ordering them from smallest to largest. However, in the fuzzy case there is no universally accepted way to do this. There are probably more than 40 methods proposed in the literature of defining $\overline{M} \leq \overline{N}$, for two fuzzy numbers $\overline{M}$ and $\overline{N}$. Here the symbol $\leq$ means "less than or equal" and not "a fuzzy subset of". A few key references on this topic are [1, 6, 7, 11, 12], where the interested reader can look up many of these methods and see their comparisons.

Here we will present only one procedure for ordering fuzzy numbers that we have used before [2, 3]. But note that different definitions of $\leq$ between fuzzy numbers can give different ordering. We first define $\leq$ between two fuzzy numbers $\overline{M}$ and $\overline{N}$. Define

$$v(\overline{M} \leq \overline{N}) = max\{min(\overline{M}(x), \overline{N}(y)) | x \leq y\}, \qquad (2.39)$$

which measures how much $\overline{M}$ is less than or equal to $\overline{N}$. We write $\overline{N} < \overline{M}$ if $v(\overline{N} \leq \overline{M}) = 1$ but $v(\overline{M} \leq \overline{N}) < \eta$, where η is some fixed fraction in $(0, 1]$. In this book we will use $\eta = 0.8$. Then $\overline{N} < \overline{M}$ if $v(\overline{N} \leq \overline{M}) = 1$ and $v(\overline{M} \leq \overline{N}) < 0.8$. We then define $\overline{M} \approx \overline{N}$ when both $\overline{N} < \overline{M}$ and $\overline{M} < \overline{N}$ are false. $\overline{M} \leq \overline{N}$ means $\overline{M} < \overline{N}$ or $\overline{M} \approx \overline{N}$. Now this $\approx$ may not be transitive. If $\overline{N} \approx \overline{M}$ and $\overline{M} \approx \overline{O}$ implies that $\overline{N} \approx \overline{O}$, then $\approx$ is transitive. However, it can happen that $\overline{N} \approx \overline{M}$ and $\overline{M} \approx \overline{O}$ but $\overline{N} < \overline{O}$ because $\overline{M}$ lies a little to the right of $\overline{N}$ and $\overline{O}$ lies a little to the right of $\overline{M}$ but $\overline{O}$ lies sufficiently far to the right of $\overline{N}$ that we obtain $\overline{N} < \overline{O}$. But this ordering is still useful in partitioning the set of fuzzy numbers up into sets $H_1, \ldots, H_K$ where [2, 3]: (1) given any $\overline{M}$ and $\overline{N}$ in H_k, $1 \leq k \leq K$, then $\overline{M} \approx \overline{N}$; and (2) given $\overline{N} \in H_i$ and $\overline{M} \in H_j$, with $i < j$, then $\overline{N} \leq \overline{M}$. Then the highest ranked fuzzy numbers lie in H_K, the second highest ranked fuzzy numbers are in H_{K-1}, etc. This result is easily seen if you graph all the fuzzy numbers on the same axis then those in H_K will be clustered together farthest to the right, proceeding from the H_K cluster to the left the next cluster will be those in H_{K-1}, etc.

There is an easy way to determine if $\overline{M} < \overline{N}$, or $\overline{M} \approx \overline{N}$, for many fuzzy numbers. First, it is easy to see that if the core of $\overline{N}$ lies completely to the right of the core of $\overline{M}$, then $v(\overline{M} \leq \overline{N}) = 1$. Also, if the core of $\overline{M}$ and the core of $\overline{N}$ overlap, then $\overline{M} \approx \overline{N}$. Now assume that the core of $\overline{N}$ lies to the right of the core of $\overline{M}$, as shown in Fig. 2.4 for triangular fuzzy numbers, and we wish to compute $v(\overline{N} \leq \overline{M})$. The value of this expression is simply y_0 in Fig. 2.4. In general, for triangular (shaped) fuzzy numbers $v(\overline{N} \leq \overline{M})$ is the height of their intersection when the core of $\overline{N}$ lies to the right of the core of $\overline{M}$. Locate η, for example $\eta = 0.8$ in this book, on the vertical

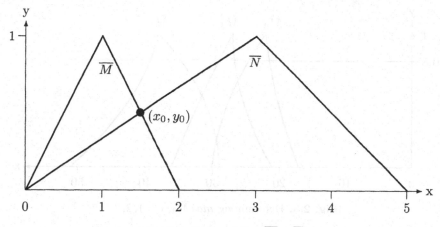

Fig. 2.4. Determining $\overline{M} < \overline{N}$

axis and then draw a horizontal line through η. If in Fig. 2.4 y_0 lies below
the horizontal line, then $\overline{M} < \overline{N}$. If y_0 lies on, or above, the horizontal line,
then $\overline{M} \approx \overline{N}$.

2.6 Optimization

Given a finite set of fuzzy numbers $\overline{M}_i$, $1 \le i \le n$, we will sometimes, in
Chaps. 9–26, want to find $max\overline{M}_i$ and/or $min\overline{M}_i$. We will use the graphical
method from Sect. 2.5. This works well when the set of fuzzy numbers is
small but becomes "messy", or too cluttered, when the set of fuzzy numbers
is large.

First consider the fuzzy numbers $\overline{M}_i$, $i = 1, 2, 3$, shown in Fig. 2.5 and we
want $max\overline{M}_i$. Clearly, using $\eta = 0.8$ from Sect. 2.5, we get $\overline{M}_2 < \overline{M}_3 < \overline{M}_1$
and $max\overline{M}_i = \overline{M}_1$.

Next consider the fuzzy numbers $\overline{M}_i$, $1 \le i \le 6$, in Fig. 2.6 and we
want $min\overline{M}_i$. Using $\eta = 0.8$ we have $\overline{M}_1 \approx \overline{M}_6 < \overline{M}_i$, $i = 2, 3, 4, 5$. Hence
$min\overline{M}_i = \{\overline{M}_1, \overline{M}_6\}$.

2.7 Discrete Versus Continuous

In Chaps. 7 and 9 through 26 there are a number of fuzzy sets we will
want to determine and they may be discrete fuzzy sets or fuzzy numbers
(now also called continuous fuzzy sets). Examples of discrete fuzzy sets may
be: (1) $\overline{LC}$ = number of lost customers per unit time; (2) $\overline{MQ}$ = maximum queue length for a certain queue in the fuzzy system; (3) $\overline{N}$ = expected (mean) number of transactions (customers, items) in the fuzzy system;

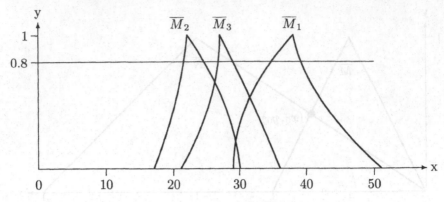

Fig. 2.5. Determining $max\overline{M}_i$, $i = 1, 2, 3$

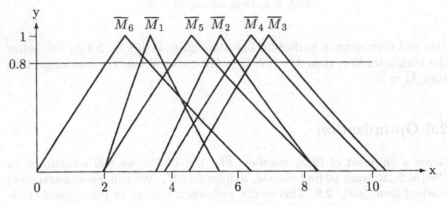

Fig. 2.6. Determining $min\overline{M}_i$, $i = 1, \ldots, 6$

(4) $\overline{S}$ = number of items sent to storage per unit time; (5) $\overline{X}$ = through-put, or number of items passing through the fuzzy system per unit time; and (6) fuzzy cost or profit $\overline{\Pi}$ when rounded off to the nearest dollar (penny, thousand dollars). We say these may be discrete fuzzy sets because it will depend on how they are computed. If the are computed as integer values they will be discrete fuzzy sets. However, they might be computed with decimals, not as integers, and then we treat them as continuous fuzzy sets. For example, since $\overline{N}$ is a mean (expected value) it will usually contain decimals and be a continuous fuzzy set. Examples of continuous fuzzy sets will be: (1) $\overline{U}$ = utilization of a server; (2) $\overline{R}$ = response time, or the expected (mean) time an item (transaction, customer) spends in the fuzzy system; and (3) $\overline{T}$ = project completion time.

In all cases we will employ simulation to approximate α-cuts of these fuzzy sets/numbers (see Chap. 7 for a justification of using simulation). We will only approximate the $\alpha = 0, 0.5, 1$ cuts of these fuzzy sets/numbers. That will give

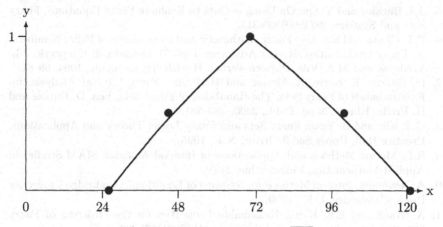

Fig. 2.7. Discrete Fuzzy Set $\overline{MQ}$

us only five points to approximate the graphs of these fuzzy sets/numbers. For an example consider approximating $\overline{MQ}$. Let $\overline{MQ}[0] = [26, 121]$, $\overline{MQ}[0.5] = [45, 100]$ and $\overline{MQ}[1] = 70$. Then we have the following five points in the graph of the membership function: $(26, 0)$, $(45, 0.5)$, $(70, 1)$, $(100, 0.5)$, $(121, 0)$. For all the discrete fuzzy sets we will now draw a continuous curve through the three points on the left side, and draw a continuous curve through the three points on the right side, producing a graph of a continuous fuzzy set. This makes the discrete fuzzy set appear to be a fuzzy number. See Fig. 2.7. The five points are plotted as small dark circles within the approximating fuzzy number.

Why did we do this to discrete fuzzy sets? This makes all our graphs uniform so the fuzzy sets may be compared using Sects. 2.5 and 2.6. Also, constructing graphs through three points is very easy using the graphing command "qbezier" in LaTeX. We will try to remind the reader about this approximating a discrete fuzzy set by a fuzzy number whenever it occurs in the rest of the book.

References

1. G. Bortolon and R. Degani: A Review of Some Methods for Ranking Fuzzy Subsets, Fuzzy Sets and Systems, 15(1985)1-19.
2. J.J. Buckley: Ranking Alternatives Using Fuzzy Numbers, Fuzzy Sets and Systems, 15(1985)21-31.
3. J.J. Buckley: Fuzzy Hierarchical Analysis, Fuzzy Sets and Systems, 17 (1985)233-247.
4. J.J. Buckley and E. Eslami: Introduction to Fuzzy Logic and Fuzzy Sets, Physica-Verlag, Heidelberg, Germany, 2002.

5. J.J. Buckley and Y. Qu: On Using α-Cuts to Evaluate Fuzzy Equations, Fuzzy Sets and Systems, 38(1990)309-312.
6. P.T. Chang and E.S. Lee: Fuzzy Arithmetic and Comparison of Fuzzy Numbers, in: Fuzzy Optimization: Recent Advances, Eds. M. Delgado, J. Kacprzyk, J.L. Verdegay and M.A. Vila, Physica-Verlag, Heidelberg, Germany, 1994, 69-81.
7. D. Dubois, E. Kerre, R. Mesiar and H. Prade: Fuzzy Interval Analysis, in: Fundamentals of Fuzzy Sets, The Handbook of Fuzzy Sets, Eds. D. Dubois and H. Prade, Kluwer Acad. Publ., 2000, 483-581.
8. G.J. Klir and B. Yuan: Fuzzy Sets and Fuzzy Logic: Theory and Applications, Prentice Hall, Upper Saddle River, N.J., 1995.
9. R.E. Moore: Methods and Applications of Interval Analysis, SIAM Studies in Applied Mathematics, Philadelphia, 1979.
10. A. Neumaier: Interval Methods for Systems of Equations, Cambridge University Press, Cambridge, U.K., 1990.
11. X. Wang and E.E. Kerre: Reasonable Properties for the Ordering of Fuzzy Quantities (I), Fuzzy Sets and Systems, 118(2001)375-385.
12. X. Wang and E.E. Kerre: Reasonable Properties for the Ordering of Fuzzy Quantities (II), Fuzzy Sets and Systems, 118(2001)387-405.

3 Fuzzy Estimation

3.1 Introduction

The first thing to do is explain how we will get fuzzy numbers , and fuzzy probabilities, from a set of confidence intervals which will be constructed from crisp data. This is done in the next two sections. Next we discuss how we can obtain fuzzy numbers for arrival rates and for service rates in queuing systems. Then we discuss the fuzzy binomial, the fuzzy normal, the fuzzy exponential and the fuzzy uniform distributions. These will be the fuzzy probability distributions used in the rest of the book. The material in this chapter has been adapted from results in references [1, 7].

3.2 Fuzzy Probabilities

Let $X = \{x_1, \ldots, x_n\}$ be a finite set and let P be a probability function defined on all subsets of X with $P(\{x_i\}) = a_i$, $1 \leq i \leq n$, $0 < a_i < 1$, all i, and $\sum_{i=1}^{n} a_i = 1$. We may substitute a fuzzy number $\overline{a}_i$ for a_i, for some i, to obtain a discrete (finite) fuzzy probability distribution . Where do these fuzzy numbers come from?

In some problems, because of the way the problem is stated, the values of all the a_i are crisp and known. For example, consider tossing a fair coin and a_1 = the probability of getting a "head" and a_2 = is the probability of obtaining a "tail". Since we assumed it to be a fair coin we must have $a_1 = a_2 = 0.5$. In this case we would not substitute a fuzzy number for a_1 or a_2. But in many other problems the a_i are not known exactly and they are either estimated from a random sample or they are obtained from "expert opinion".

Suppose we have the results of a random sample to estimate the value of a_1. We would construct a set of confidence intervals for a_1 and then put these together to get the fuzzy number $\overline{a}_1$ for a_1. This method of building a fuzzy number from confidence intervals is discussed in detail in the next section.

Assume that we do not know the values of the a_i and we do not have any data to estimate their values. Then we may obtain numbers for the a_i from some group of experts. This group could consist of only one expert. This case includes subjective, or "personal", probabilities. First assume we have

James J. Buckley: *Simulating Fuzzy Systems*, StudFuzz **171**, 19–35 (2005)
www.springerlink.com

only one expert and he is to estimate the value of some probability p. We can solicit this estimate from the expert as is done in estimating job times in project scheduling ([11], Chap. 13). Let a = the "pessimistic" value of p, or the smallest possible value, let c = be the "optimistic" value of p, or the highest possible value, and let b = the most likely value of p. We then ask the expert to give values for a, b, c and we construct the triangular fuzzy number $\overline{p} = (a/b/c)$ for p. If we have a group of N experts all to estimate the value of p we solicit the a_i, b_i and c_i, $1 \le i \le N$, from them. Let a be the average of the a_i, b is the mean of the b_i and c is the average of the c_i. The simplest thing to do is to use $(a/b/c)$ for $\overline{p}$.

3.3 Fuzzy Numbers from Confidence Intervals

We will be using fuzzy numbers for parameters in probability density functions (probability mass functions, the discrete case) and in this section we show how we obtain these fuzzy numbers from a set of confidence intervals. Let X be a random variable with probability density function (or probability mass function) $f(x; \theta)$ for single parameter θ. It is easy to generalize our method to the case where θ is a vector of parameters. Assume that θ is unknown and it must be estimated from a random sample $X_1, \ldots, X_n$. Let $Y = u(X_1, \ldots, X_n)$ be a statistic used to estimate θ. Given the values of these random variables $X_i = x_i$, $1 \le i \le n$, we obtain a point estimate $\theta^* = y = u(x_1, \ldots, x_n)$ for θ. We would never expect this point estimate to exactly equal θ so we often also compute a $(1 - \beta)100\%$ confidence interval for θ. We are using β here since α, usually employed for confidence interval, is reserved for α-cuts of fuzzy numbers. In this confidence interval one usually sets β equal to 0.10, 0.05 or 0.01.

We propose to find the $(1-\beta)100\%$ confidence interval for all $0.01 \le \beta < 1$. Starting at 0.01 is arbitrary and you could begin at 0.001 or 0.005 etc. Denote these confidence intervals as

$$[\theta_1(\beta), \theta_2(\beta)] , \tag{3.1}$$

for $0.01 \le \beta < 1$. Add to this the interval $[\theta^*, \theta^*]$ for the 0% confidence interval for θ. Then we have $(1 - \beta)100\%$ confidence interval for θ for $0.01 \le \beta \le 1$.

Now place these confidence intervals, one on top of the other, to produce a triangular shaped fuzzy number $\overline{\theta}$ whose α-cuts are the confidence intervals. We have

$$\overline{\theta}[\alpha] = [\theta_1(\alpha), \theta_2(\alpha)] , \tag{3.2}$$

for $0.01 \le \alpha \le 1$. All that is needed is to finish the "bottom" of $\overline{\theta}$ to make it a complete fuzzy number. We will simply drop the graph of $\overline{\theta}$ straight down to complete its α-cuts so

$$\overline{\theta}[\alpha] = [\theta_1(0.01), \theta_2(0.01)] \,, \tag{3.3}$$

for $0 \le \alpha < 0.01$. In this way we are using more information in $\overline{\theta}$ than just a point estimate, or just a single interval estimate.

The following example shows that the fuzzy mean of the normal probability density will be a triangular shaped fuzzy number. However, for simplicity, throughout this book we will always use triangular fuzzy numbers for the fuzzy values of uncertain parameters in probability density (mass) functions.

Example 3.3.1

Consider X a random variable with probability density function $N(\mu, 100)$, which is the normal probability density with unknown mean μ and known variance $\sigma^2 = 100$. To estimate μ we obtain a random sample $X_1, \ldots, X_n$ from $N(\mu, 100)$. Suppose the mean of this random sample turns out to be 28.6. Then a $(1 - \beta)100\%$ confidence interval for μ is

$$[\theta_1(\beta), \theta_2(\beta)] = [28.6 - z_{\beta/2}10/\sqrt{n}, 28.6 + z_{\beta/2}10/\sqrt{n}] \,, \tag{3.4}$$

where $z_{\beta/2}$ is defined as

$$\int_{-\infty}^{z_{\beta/2}} N(0,1)dx = 1 - \beta/2 \,, \tag{3.5}$$

and $N(0,1)$ denotes the normal density with mean zero and unit variance. To obtain a graph of fuzzy μ, or $\overline{\mu}$, let $n = 64$. Then we evaluated (3.4) and (3.5) using Maple [9] and the final graph of $\overline{\mu}$ is shown in Fig. 3.1, without dropping the graph straight down to the x-axis at the end points.

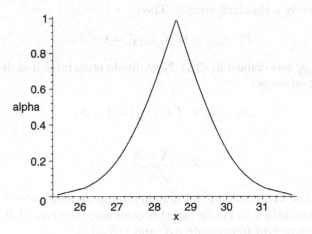

Fig. 3.1. Fuzzy Mean $\overline{\mu}$ in Example 3.3.1

In future chapters we will have fuzzy numbers for the parameters in the probability density (mass) functions producing fuzzy probability density (mass) functions. These fuzzy probability density (mass) functions are discussed in the next chapter.

3.4 Fuzzy Arrival/Service Rates

In this section we concentrate on deriving fuzzy numbers for the arrival rate, and the service rate, in a queuing system. We consider the fuzzy arrival rate first.

3.4.1 Fuzzy Arrival Rate

We assume that we have Poisson arrivals ([11], Chap. 15) which means that there is a positive constant λ so that the probability of k arrivals per unit time is

$$\lambda^k \exp(-\lambda)/k! \,, \tag{3.6}$$

the Poisson probability function. We need to estimate λ, the arrival rate, so we take a random sample $X_1, \ldots, X_m$ of size m. In the random sample X_i is the number of arrivals per unit time, in the ith observation. Let S be the sum of the X_i and let $\overline{X}$ be S/m. Here, $\overline{X}$ is not a fuzzy set but the mean.

Now S is Poisson with parameter $m\lambda$ ([8], p. 298). Assuming that $m\lambda$ is sufficiently large (say, at least 30), we may use the normal approximation ([8], p. 317), so the statistic

$$W = \frac{S - m\lambda}{\sqrt{m\lambda}} \,, \tag{3.7}$$

is approximately a standard normal. Then

$$P[-z_{\beta/2} < W < z_{\beta/2}] = 1 - \beta \,, \tag{3.8}$$

where the $z_{\beta/2}$ was defined in (3.5). Now divide numerator and denominator of W by m and we get

$$P[-z_{\beta/2} < Z < z_{\beta/2}] = 1 - \beta \,, \tag{3.9}$$

where

$$Z = \frac{\overline{X} - \lambda}{\sqrt{\lambda/m}} \,. \tag{3.10}$$

From these last two equations we may derive an approximate $(1 - \beta)100\%$ confidence interval for λ. Let us call this confidence interval $[l(\beta), r(\beta)]$.

We now show how to compute $l(\beta)$ and $r(\beta)$. Let

$$f(\lambda) = \sqrt{m}(\overline{X} - \lambda)/\sqrt{\lambda}\,. \tag{3.11}$$

Now $f(\lambda)$ has the following properties: (1) it is strictly decreasing for $\lambda > 0$; (2) it is zero for $\lambda > 0$ only at $\overline{X} = \lambda$; (3) the limit of f, as λ goes to ∞ is $-\infty$; and (4) the limit of f as λ approaches zero from the right is ∞. Hence, (1) the equation $z_{\beta/2} = f(\lambda)$ has a unique solution $\lambda = l(\beta)$; and (2) the equation $-z_{\beta/2} = f(\lambda)$ also has a unique solution $\lambda = r(\beta)$.

We may find these unique solutions. Let

$$V = \sqrt{z_{\beta/2}^2/m + 4\overline{X}}\,, \tag{3.12}$$

$$z_1 = \left[-\frac{z_{\beta/2}}{\sqrt{m}} + V\right]\bigg/2\,, \tag{3.13}$$

and

$$z_2 = \left[\frac{z_{\beta/2}}{\sqrt{m}} + V\right]\bigg/2\,. \tag{3.14}$$

Then $l(\beta) = z_1^2$ and $r(\beta) = z_2^2$.

We now substitute α for β to get the α-cuts of fuzzy number $\overline{\lambda}$. Add the point estimate, when $\alpha = 1$, $\overline{X}$, for the 0% confidence interval. Now as α goes from 0.01 (99% confidence interval) to one (0% confidence interval) we get the fuzzy number for λ. As before, we drop the graph straight down at the ends to obtain a complete fuzzy number.

Example 3.4.1.1

Suppose $m = 100$ and we obtained $\overline{X} = 25$. We evaluated (3.12) through (3.14) using Maple [9] and then the graph of $\overline{\lambda}$ is shown in Fig. 3.2, without dropping the graph straight down to the x-axis at the end points. However, in the rest of the book we will use a triangular fuzzy number for $\overline{\lambda}$.

3.4.2 Fuzzy Service Rate

Let μ be the average (expected) service rate, in the number of service completions per unit time, for a busy server. Then $1/\mu$ is the average (expected) service time. The probability density of the time interval between successive service completions is ([11], Chap. 15)

$$(1/\mu)\exp(-t/\mu)\,, \tag{3.15}$$

for $t > 0$, the exponential probability density function. Let $X_1, \ldots, X_n$ be a random sample from this exponential density function. Then the maximum likelihood estimator for μ is $\overline{X}$ ([8], p. 344), the mean of the random sample (not a fuzzy set). We know that the probability density for $\overline{X}$ is the gamma

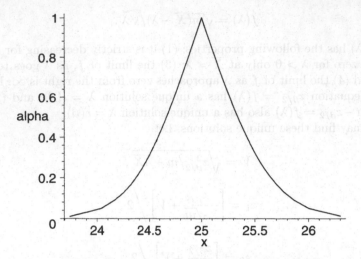

Fig. 3.2. Fuzzy Arrival Rate $\overline{\lambda}$ in Example 3.4.1.1

([8], p. 297) with mean μ and variance μ^2/n ([8], p. 351). If n is sufficiently large we may use the normal approximation to determine approximate confidence intervals for μ. Let

$$Z = (\sqrt{n}[\overline{X} - \mu])/\mu , \qquad (3.16)$$

which is approximately normally distributed with zero mean and unit variance, provided n is sufficiently large. See Fig. 3.2 in [8] for $n = 100$ which shows the approximation is quite good if $n = 100$. The graph in Fig. 3.2 in [8] is for the chi-square distribution which is a special case of the gamma distribution. So we now assume that $n \geq 100$ and use the normal approximation to the gamma.

An approximate $(1 - \beta)100\%$ confidence interval for μ is obtained from

$$P[-z_{\beta/2} < Z < z_{\beta/2}] = 1 - \beta , \qquad (3.17)$$

where β was defined in (3.5). After solving for μ we get

$$P[L(\beta) < \mu < R(\beta)] = 1 - \beta , \qquad (3.18)$$

where

$$L(\beta) = [\sqrt{n}\,\overline{X}]/[z_{\beta/2} + \sqrt{n}] , \qquad (3.19)$$

and

$$R(\beta) = [\sqrt{n}\,\overline{X}]/[\sqrt{n} - z_{\beta/2}] . \qquad (3.20)$$

An approximate $(1 - \beta)100\%$ confidence interval for μ is

$$\left[\frac{\sqrt{n}\,\overline{X}}{z_{\beta/2} + \sqrt{n}}, \frac{\sqrt{n}\,\overline{X}}{\sqrt{n} - z_{\beta/2}} \right] . \qquad (3.21)$$

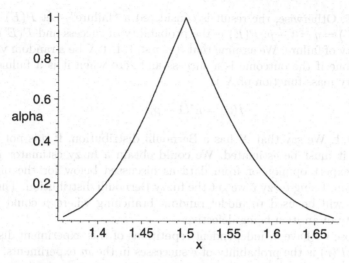

Fig. 3.3. Fuzzy Service Rate $\overline{\mu}$ in Example 3.4.2.1

Example 3.4.2.1

If $n = 400$ and $\overline{X} = 1.5$, then we get

$$\left[\frac{30}{z_{\beta/2} + 20}, \frac{30}{20 - z_{\beta/2}}\right], \tag{3.22}$$

for a $(1 - \beta)100\%$ confidence interval for the service rate μ. Now we can put these confidence intervals together, one on top of another, to obtain a fuzzy number $\overline{\mu}$ for the service rate. We evaluated (3.22) using Maple [9] for $0.01 \leq \beta \leq 1$ and the graph of the fuzzy service rate, without dropping the graph straight down to the x-axis at the end points, is in Fig. 3.3. For simplicity we use triangular fuzzy numbers for $\overline{\mu}$ in the rest of the book.

3.5 Fuzzy Probability Distributions

Now we can introduce the fuzzy binomial, the fuzzy normal, the fuzzy exponential and the fuzzy uniform distributions. How we compute fuzzy probabilities with these fuzzy distributions is discussed in the next chapter. These fuzzy probability distributions may be used to model random branching or arrival/service times.

3.5.1 Fuzzy Binomial

Let $X = \{x_1, \ldots, x_n\}$ and let E be a non-empty, proper, subset of X. We have an experiment where the result is considered a "success" if the outcome

x_i is in E. Otherwise, the result is considered a "failure". Let $P(E) = p$ so that $P(E') = q = 1 - p$. $P(E)$ is the probability of success and $P(E')$ is the probability of failure. We assume that $0 < p < 1$. Let X be a random variable which is one if the outcome is a success and zero when it is a failure. The probability mass function of X is

$$f(x) = p^x(1 - p)^{1-x},\tag{3.23}$$

for $x = 0, 1$. We say that X has a Bernoulli distribution. If p is not known precisely it must be estimated. We could obtain a fuzzy estimator $\bar{p}$ for p through expert opinion or from data as discussed below for the binomial distribution. Using fuzzy $\bar{p}$ we get the fuzzy Bernoulli distribution. The fuzzy Bernoulli will be used to model random branching where p could be the probability of an item being defective.

Suppose we have m independent repetitions of this experiment discussed above. If $P(r)$ is the probability of r successes in the m experiments, then

$$P(r) = \binom{m}{r} p^r q^{m-r},\tag{3.24}$$

for $r = 0, 1, 2, \ldots, m$, gives the binomial distribution.

In these experiments let us assume that $P(E) = p$ is not known precisely and it needs to be estimated from data, or obtained from expert opinion. Let us assume that we estimate p from some crisp data.

Example 3.5.1.1

Suppose we observe the system during $N = 500$ independent repetitions and find that there have been 100 "successes". Using the normal approximation to the binomial an approximate $(1 - \beta)100\%$ confidence interval for p is

$$\left[0.2 - z_{\beta/2}\sqrt{\frac{0.2(1 - 0.2)}{500}}, 0.2 + z_{\beta/2}\sqrt{\frac{0.2(1 - 0.2)}{500}}\right],\tag{3.25}$$

where $z_{\beta/2}$ is defined in (3.5). We evaluated (3.25) using Maple [9] and then the graph of $\bar{p}$ is shown in Fig. 3.4, without dropping the graph straight down to the x-axis at the end points. However, in further calculations we will be using triangular fuzzy numbers for the $\bar{p}$.

So, we substitute $\bar{p}$ for p and $\bar{q} = 1 - \bar{p}$ for q, to obtain the fuzzy binomial distribution which is discussed further in the next chapter.

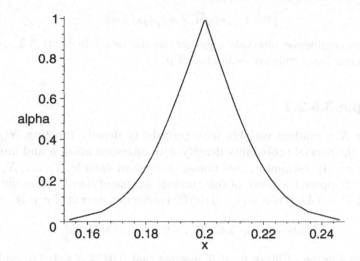

Fig. 3.4. Fuzzy Probability $\bar{p}$ in Example 3.5.1.1

3.5.2 Fuzzy Estimator of μ in the Normal

Consider X a random variable with probability density function $N(\mu, \sigma^2)$, which is the normal probability density with unknown mean μ and unknown variance σ^2. To estimate μ we obtain a random sample $X_1, \ldots, X_n$ from $N(\mu, \sigma^2)$. Suppose the mean of this random sample turns out to be $\bar{x}$, which is a crisp number, not a fuzzy number. Also, let s^2 be the sample variance. Our point estimator of μ is $\bar{x}$. If the values of the random sample are $x_1, \ldots, x_n$ then the expression we will use for s^2 in this book is

$$s^2 = \sum_{i=1}^{n} (x_i - \bar{x})^2 / (n-1) . \tag{3.26}$$

We will use this form of s^2, with denominator $(n-1)$, so that it is an unbiased estimator of σ^2.

It is known that $(\bar{x} - \mu)/(s/\sqrt{n})$ has a (Student's) t distribution with $n-1$ degrees of freedom (Sect. 7.2 of [8]). It follows that

$$P\left(-t_{\beta/2} \leq \frac{\bar{x} - \mu}{s/\sqrt{n}} \leq t_{\beta/2}\right) = 1 - \beta , \tag{3.27}$$

where $t_{\beta/2}$ is defined from the (Student's) t distribution, with $n-1$ degrees of freedom, so that the probability of exceeding it is $\beta/2$. Now solve the inequality for μ giving

$$P\left(\bar{x} - t_{\beta/2}s/\sqrt{n} \leq \mu \leq \bar{x} + t_{\beta/2}s/\sqrt{n}\right) = 1 - \beta . \tag{3.28}$$

For this we immediately obtain the $(1 - \beta)100\%$ confidence interval for μ

$$\left[\bar{x} - t_{\beta/2}s/\sqrt{n}, \bar{x} + t_{\beta/2}s/\sqrt{n}\right] .$$ (3.29)

Put these confidence intervals together, as discussed in Sect. 3.3, and we obtain $\bar{\mu}$ our fuzzy number estimator of μ.

Example 3.5.2.1

Consider X a random variable with probability density function $N(\mu, \sigma^2)$, which is the normal probability density with unknown mean μ and unknown variance σ^2. To estimate μ we obtain a random sample $X_1, \ldots, X_n$ from $N(\mu, \sigma^2)$. Suppose the mean of this random sample of size 25 turns out to be 28.6 and $s^2 = 3.42$. Then a $(1 - \beta)100\%$ confidence interval for μ is

$$\left[28.6 - t_{\beta/2}\sqrt{3.42/25}, 28.6 + t_{\beta/2}\sqrt{3.42/25}\right] .$$ (3.30)

To obtain a graph of fuzzy μ, or $\bar{\mu}$, assume that $0.01 \le \beta \le 1$. We evaluated (3.30) using Maple [9] and then the graph of $\bar{\mu}$ is shown in Fig. 3.5, without dropping the graph straight down to the x-axis at the end points.

To complete the picture we draw short vertical line segments, from the horizontal axis up to the graph, at the end points of the base of the fuzzy number $\bar{\mu}$. The base ($\bar{\mu}[0]$) is a 99% confidence interval for μ.

3.5.3 Fuzzy Estimator of σ^2 in the Normal

We first construct a fuzzy estimator for σ^2 using the usual confidence intervals for the variance from a normal distribution and we show this fuzzy estimator is biased. Then we construct an unbiased fuzzy estimator for the variance.

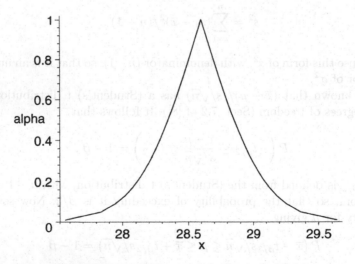

Fig. 3.5. Fuzzy Estimator $\bar{\mu}$ in Example 3.5.2.1

Biased Fuzzy Estimator

Consider X a random variable with probability density function $N(\mu, \sigma^2)$, which is the normal probability density with unknown mean μ and unknown variance σ^2. To estimate σ^2 we obtain a random sample $X_1, \ldots, X_n$ from $N(\mu, \sigma^2)$. Our point estimator for the variance will be s^2.

We know that (Sect. 7.4 in [8]) $(n-1)s^2/\sigma^2$ has a chi-square distribution with $n-1$ degrees of freedom. Then

$$P\left(\chi^2_{L,\beta/2} \leq (n-1)s^2/\sigma^2 \leq \chi^2_{R,\beta/2}\right) = 1 - \beta \, , \tag{3.31}$$

where $\chi^2_{R,\beta/2}$ $(\chi^2_{L,\beta/2})$ is the point on the right (left) side of the χ^2 density where the probability of exceeding (being less than) it is $\beta/2$. The χ^2 distribution has $n-1$ degrees of freedom. Solve the inequality for σ^2 and we see that

$$P\left(\frac{(n-1)s^2}{\chi^2_{R,\beta/2}} \leq \sigma^2 \leq \frac{(n-1)s^2}{\chi^2_{L,\beta/2}}\right) = 1 - \beta \, . \tag{3.32}$$

From this we obtain the usual $(1 - \beta)100\%$ confidence intervals for σ^2

$$\left[(n-1)s^2/\chi^2_{R,\beta/2}, (n-1)s^2/\chi^2_{L,\beta/2}\right] \, . \tag{3.33}$$

Put these confidence intervals together, as discussed in Sect. 3.3, and we obtain $\overline{\sigma}^2$ our fuzzy number estimator of σ^2.

We now show that this fuzzy estimator is biased because the vertex of the triangular shaped fuzzy number $\overline{\sigma}^2$, where the membership value equals one, is not at s^2. We say a fuzzy estimator is biased when its vertex is not at the point (crisp) estimator. We obtain the vertex of $\overline{\sigma}^2$ when $\beta = 1.0$. Let

$$factor = \frac{n-1}{\chi^2_{R,0.50}} = \frac{n-1}{\chi^2_{L,0.50}} \, , \tag{3.34}$$

after we substitute $\beta = 1$. Then the 0% confidence interval for the variance is

$$[(factor)(s^2), (factor)(s^2)] = (factor)(s^2) \, . \tag{3.35}$$

Since $factor \neq 1$ the fuzzy number $\overline{\sigma}^2$ is not centered at s^2. Table 3.1 shows some values of $factor$ for various choices for n. We see that $factor \to 1$ as $n \to \infty$ but $factor$ is substantially larger than one for small values on n. This fuzzy estimator is biased and we will construct an unbiased (vertex at s^2) in the next subsection.

Unbiased Fuzzy Estimator

In deriving the usual confidence interval for the variance we start with recognizing that $(n-1)s^2/\sigma^2$ has a χ^2 distribution with $n-1$ degrees of freedom. Then for a $(1 - \beta)100\%$ confidence interval we may find a and b so that

Table 3.1. Values of $factor$ for Various Values of n

n	$factor$
10	1.0788
20	1.0361
50	1.0138
100	1.0068
500	1.0013
1000	1.0007

$$P\left(a \le \frac{(n-1)s^2}{\sigma^2} \le b\right) = 1 - \beta . \tag{3.36}$$

The usual confidence interval has a and b so that the probabilities in the "two tails" are equal. That is, $a = \chi^2_{L,\beta/2}$ ($b = \chi^2_{R,\beta/2}$) so that the probability of being less (greater) than a (b) is $\beta/2$. But we do not have to pick the a and b this way ([8], p. 378). We will change the way we pick the a and b so that the fuzzy estimator is unbiased.

Assume that $0.01 \le \beta \le 1$. Now this interval for β is fixed and also n and s^2 are fixed. Define

$$L(\lambda) = [1 - \lambda]\chi^2_{R,0.005} + \lambda(n-1) , \tag{3.37}$$

and

$$R(\lambda) = [1 - \lambda]\chi^2_{L,0.005} + \lambda(n-1) . \tag{3.38}$$

Then a confidence interval for the variance is

$$\left[\frac{(n-1)s^2}{L(\lambda)}, \frac{(n-1)s^2}{R(\lambda)}\right] , \tag{3.39}$$

for $0 \le \lambda \le 1$. We start with a 99% confidence interval when $\lambda = 0$ and end up with a 0% confidence interval for $\lambda = 1$. Notice that now the 0% confidence interval is $[s^2, s^2] = s^2$ and it is unbiased. As usual, we place these confidence intervals one on top of another to obtain our (unbiased) fuzzy estimator $\overline{\sigma}^2$ for the variance. Our confidence interval for σ, the population standard deviation, is

$$\left[\sqrt{(n-1)/L(\lambda)}s, \sqrt{(n-1)/R(\lambda)}s\right] . \tag{3.40}$$

Let us compare the methods in this subsection to those in the previous subsection. Let χ^2 be the chi-square probability density with $n-1$ degrees of freedom. The mean of χ^2 is $n-1$ and the median is the point md where $P(X \le md) = P(X \ge md) = 0.5$. We assume β is in the interval $[0.01, 1]$. In the previous subsection as β continuously increases from 0.01 to 1, $\chi^2_{L,\beta/2}$ ($\chi^2_{R,\beta/2}$) starts at $\chi^2_{L,0.005}$ ($\chi^2_{R,0.005}$) and increases (decreases) to $\chi^2_{L,0.5}$ ($\chi^2_{R,0.5}$) which equals to the median. Recall that $\chi^2_{L,\beta/2}$ ($\chi^2_{R,\beta/2}$) is the point on the χ^2 density where the probability of being less (greater) that it equals $\beta/2$. From Table 3.1 we see that the median is always less than $n-1$. This produces the bias in the fuzzy estimator in that subsection. In this subsection as λ continuously increases from zero to one $L(\lambda)$ ($R(\lambda)$) decreases (increases) from $\chi^2_{R,0.005}$ ($\chi^2_{L,0.005}$) to $n-1$. At $\lambda = 1$ we get $L(1) = R(1) = n-1$ and the vertex (membership value one) is at s^2 and it is now unbiased.

We will use this fuzzy estimator $\overline{\sigma}^2$ constructed in this section for σ^2 in the rest of this book. Given a value of $\lambda = \lambda^* \in [0,1]$ one may wonder what is the corresponding value of β for the confidence interval. We now show how to get the β. Let $L^* = L(\lambda^*)$ and $R^* = R(\lambda^*)$. Define

$$l = \int_0^{R^*} \chi^2 dx , \qquad (3.41)$$

and

$$r = \int_{L^*}^{\infty} \chi^2 dx , \qquad (3.42)$$

and then $\beta = l + r$. Now l (r) need not equal $\beta/2$. Both of these integrals above are easily evaluated using Maple [9]. The chi-square density inside these integrals has $n-1$ degrees of freedom.

Example 3.5.3.1

Consider X a random variable with probability density function $N(\mu, \sigma^2)$, which is the normal probability density with mean μ and unknown variance σ^2. To estimate σ^2 we obtain a random sample $X_1, \ldots, X_n$ from $N(\mu, \sigma^2)$. Suppose $n = 25$ and we calculate $s^2 = 3.42$. Then a confidence interval for σ^2 is

$$\left[\frac{82.08}{L(\lambda)}, \frac{82.08}{R(\lambda)} \right] . \qquad (3.43)$$

To obtain a graph of fuzzy σ^2, or $\overline{\sigma}^2$, first assume that $0.01 \leq \beta \leq 1$. We evaluated (3.43) using Maple [9] and then the graph of $\overline{\sigma}^2$ is shown in Fig. 3.6, without dropping the graph straight down to the x-axis at the end points.

To complete the picture we draw short vertical line segments, from the horizontal axis up to the graph, at the end points of the base of the fuzzy number $\overline{\sigma}^2$. The base ($\overline{\sigma}^2[0]$) is a 99% confidence interval for σ^2.

To complete this subsection let us present one graph of our fuzzy estimator $\overline{\sigma}$ of σ. Alpha-cuts of $\overline{\sigma}$ are

$$\left[\frac{9.06}{\sqrt{L(\lambda)}}, \frac{9.06}{\sqrt{R(\lambda)}} \right] . \qquad (3.44)$$

Assuming $0.01 \leq \beta \leq 1$, the graph is in Fig. 3.7.

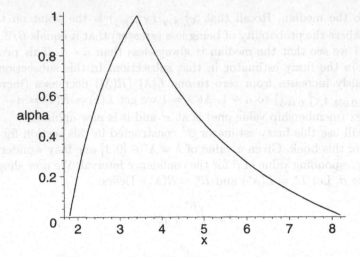

Fig. 3.6. Fuzzy Estimator $\overline{\sigma}^2$ in Example 3.5.3.1

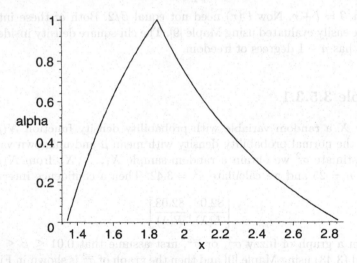

Fig. 3.7. Fuzzy Estimator $\overline{\sigma}$ in Example 3.5.3.1

3.5.4 Fuzzy Exponential

The exponential has density $f(x; \lambda) = \lambda \exp(-\lambda x)$ for $x \geq 0$ and $f(x; \lambda) = 0$ otherwise, where $\lambda > 0$. The mean and variance of exponential are $1/\lambda$ and $1/\lambda^2$, respectively. Now consider the exponential for fuzzy number $\overline{\lambda} > 0$. We discussed constructing $\overline{\lambda}$ in Sect. 3.4.1. The exponential density with fuzzy $\overline{\lambda}$ is the fuzzy exponential density. More on the fuzzy exponential in the next chapter.

3.5.5 Fuzzy Uniform

Let $U(a,b)$, for $0 \leq a < b$, denote the uniform distribution whose density function is $f(x) = 1/(b-a)$ for $a < x < b$, and $f(x) = 0$ otherwise. The uniform distribution will be used to model times between arrivals in a queuing system and/or service times in a server.

Assume that we do not know a and b precisely so they must be estimated. Let $X_1, \ldots, X_n$ be a random sample from $U(a,b)$ and let $x_1, \ldots, x_n$ be the values of this random sample. Define $x_{min} = min\{x_i | 1 \leq i \leq n\}$ and $x_{max} = max\{x_i | 1 \leq i \leq n\}$. Then x_{min} (x_{max}) is our point estimator of a (b). We may now construct $(1-\beta)100\%$ confidence intervals for a (b), for $0.01 \leq \beta \leq 1$, and placing these intervals one on top of another (with completing the base as described above) to obtain our fuzzy estimator $\overline{a}$ ($\overline{b}$) of a (b). We will not describe the mechanics of building these confidence intervals in this book because to do so would divert us into too much theoretical work in probability and statistics. The solution for the confidence intervals is complicated [10] and not to be found in statistics books. However, we do show the results in Figs. 3.8 and 3.9. Both figures give $\overline{a}$ and $\overline{b}$ and in Fig. 3.8 the left side of $\overline{a}$ was cut off because a is non-negative. These graphs were done using Maple [9].

In some applications (Chaps. 12, 14 and 19) we will use triangular fuzzy numbers for $\overline{a}$ and $\overline{b}$. This is not correct since the $\overline{a}$ and $\overline{b}$ in Figs. 3.8 and 3.9 are not triangular fuzzy numbers. It is just a habit of the author to use triangular fuzzy numbers in applications. However, changing the $\overline{a}$ and $\overline{b}$ in these chapters to look like those in Figs. 3.8 and 3.9 would probably only cause small changes in the results.

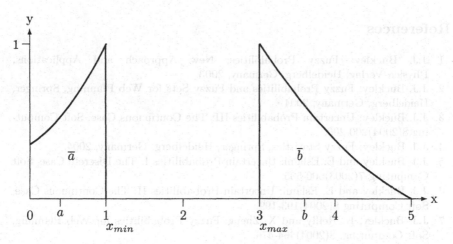

Fig. 3.8. Fuzzy Estimators $\overline{a}$ and $\overline{b}$ in $U(a,b)$

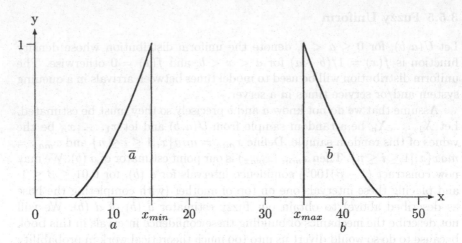

Fig. 3.9. Fuzzy Estimators $\bar{a}$ and $\bar{b}$ in the Uniform Distribution

3.6 Summary

We saw in this chapter that our fuzzy estimators will be triangular shaped fuzzy numbers, where we complete the base by drawing short vertical line segments from the horizontal axis up to the graph, and the base represents a 99% confidence interval. In the rest of this book, for simplicity, our fuzzy estimators will be triangular fuzzy numbers whose base (support) is the 99% confidence interval.

References

1. J.J. Buckley: Fuzzy Probabilities: New Approach and Applications, Physica-Verlag, Heidelberg, Germany, 2003.
2. J.J. Buckley: Fuzzy Probabilities and Fuzzy Sets for Web Planning, Springer, Heidelberg, Germany, 2004.
3. J.J. Buckley: Uncertain Probabilities III: The Continuous Case, Soft Computing 8(2004)200-206.
4. J.J. Buckley: Fuzzy Statistics, Springer, Heidelberg, Germany, 2004.
5. J.J. Buckley and E. Eslami: Uncertain Probabilities I: The Discrete Case, Soft Computing 7(2003)500-505.
6. J.J. Buckley and E. Eslami: Uncertain Probabilities II: The Continuous Case, Soft Computing 8(2004)193-199.
7. J.J. Buckley, K. Reilly and X. Zheng: Fuzzy Probabilities for Web Planning, Soft Computing, 8(2004)464-476.
8. R.V. Hogg and E.A. Tanis: Probability and Statistical Inference, Sixth Edition, Prentice Hall, Upper Saddle River, N.J., 2001.

9. Maple 9, Waterloo Maple Inc., Waterloo, Canada.
10. Personal Communication from Professor N. Chernov, Mathematics Department, UAB, Birmingham, Alabama.
11. H.A. Taha: Operations Research, Fifth Edition, Macmillan, N.Y., 1992.

4 Fuzzy Probability Theory

4.1 Introduction

In this chapter we look more closely at the fuzzy binomial distribution, the fuzzy Poisson, and at the fuzzy normal, exponential and uniform densities. We need to be familiar with these fuzzy probability functions because their use transforms crisp systems theory to fuzzy system theory (Chap. 5). The material in this chapter is based on [1–4].

4.2 Fuzzy Binomial

As before $X = \{x_1, \ldots, x_n\}$ and let E be a non-empty, proper, subset of X. We have an experiment where the result is considered a "success" if the outcome x_i is in E. Otherwise, the result is considered a "failure". Let $P(E) = p$ so that $P(E') = q = 1 - p$. $P(E)$ is the probability of success and $P(E')$ is the probability of failure. We assume that $0 < p < 1$.

Suppose we have m independent repetitions of this experiment. If $P(r)$ is the probability of r successes in the m experiments, then

$$P(r) = \binom{m}{r} p^r q^{m-r} , \tag{4.1}$$

for $r = 0, 1, 2, \ldots, m$, gives the binomial distribution.

In these experiments let us assume that $P(E)$ is not known precisely and it needs to be estimated, or obtained from expert opinion. So the p value is uncertain and we substitute $\overline{p}$ for p (see Sect. 3.5.1) and $\overline{q} = 1 - \overline{p}$. Now let $\overline{P}(r)$ be the fuzzy probability of r successes in m independent trials of the experiment. Then we obtain

$$\overline{P}(r)[\alpha] = \left\{ \binom{m}{r} p^r q^{m-r} \Big| \mathbf{S} \right\} , \tag{4.2}$$

for $0 \le \alpha \le 1$, where $\mathbf{S}$ is the statement "$p \in \overline{p}[\alpha], q \in \overline{q}[\alpha], p + q = 1$". There is uncertainty in the value of p and q but there is no uncertainty in the fact that the sum of p and q must equal one. Notice that $\overline{P}(r)$ is not $\binom{m}{r} \overline{p}^r \overline{q}^{m-r}$. If $\overline{P}(r)[\alpha] = [P_{r1}(\alpha), P_{r2}(\alpha)]$, then

James J. Buckley: *Simulating Fuzzy Systems*, StudFuzz **171**, 37–48 (2005)
www.springerlink.com © Springer-Verlag Berlin Heidelberg 2005

$$P_{r1}(\alpha) = min\left\{\binom{m}{r}p^r q^{m-r}|\mathbf{S}\right\} , \qquad (4.3)$$

and

$$P_{r2}(\alpha) = max\left\{\binom{m}{r}p^r q^{m-r}|\mathbf{S}\right\} . \qquad (4.4)$$

Example 4.2.1

Let $p \approx 0.4$, $q \approx 0.6$ and $m = 3$. Since p and q are uncertain we use $\overline{p} = (0.3/0.4/0.5)$ for p and $\overline{q} = (0.5/0.6/0.7)$ for q. Now we will calculate the fuzzy number $\overline{P}(2)$. If $p \in \overline{p}[\alpha]$ then $q = 1 - p \in \overline{q}[\alpha]$. Equations (4.3) and (4.4) become

$$P_{r1}(\alpha) = min\{3p^2 q|\mathbf{S}\} , \qquad (4.5)$$

and

$$P_{r2}(\alpha) = max\{3p^2 q|\mathbf{S}\} . \qquad (4.6)$$

Since $d(3p^2(1 - p))/dp > 0$ on $\overline{p}[0]$ we obtain

$$\overline{P}(2)[\alpha] = [3(p_1(\alpha))^2(1 - p_1(\alpha)), 3(p_2(\alpha))^2(1 - p_2(\alpha))] , \qquad (4.7)$$

where $\overline{p}[\alpha] = [p_1(\alpha), p_2(\alpha)] = [0.3 + 0.1\alpha, 0.5 - 0.1\alpha]$.

We now determine the fuzzy and the fuzzy variance of the fuzzy binomial distribution. In the crisp case we know $\mu = mp$ and $\sigma^2 = mpq$. Does $\overline{\mu} = m\overline{p}$ and $\overline{\sigma}^2 = m\overline{p} \cdot \overline{q}$? We now argue that the first result is correct but the second is not correct. We see that

$$\overline{\mu}[\alpha] = \left\{\sum_{r=0}^m r\binom{m}{r}p^r q^{m-r}|\mathbf{S}\right\} , \qquad (4.8)$$

which simplifies to

$$\overline{\mu}[\alpha] = \{mp|\mathbf{S}\} . \qquad (4.9)$$

Since $q = 1 - p$ and $\overline{q} = 1 - \overline{p}$ we may always find a $q \in \overline{q}[\alpha]$, given a $p \in \overline{p}[\alpha]$, so that $p + q = 1$. Hence, we may reduce $\mathbf{S}$ to simply "$p \in \overline{p}[\alpha]$". It follows that $\overline{\mu}[\alpha] = m\overline{p}[\alpha]$ and then $\overline{\mu} = m\overline{p}$.

We now argue that $\overline{\sigma}^2 \leq m\overline{p} \cdot \overline{q}$ and they may not be equal. We see first, as we get (4.9) from (4.8), that

$$\overline{\sigma}^2[\alpha] = \{mpq|\mathbf{S}\} . \qquad (4.10)$$

We now argue that $\overline{\sigma}^2[\alpha] \subset (m\overline{p}\overline{q})[\alpha]$, $0 \leq \alpha \leq 1$, which implies $\overline{\sigma}^2 \leq m\overline{p}\overline{q}$. Let $s \in \overline{\sigma}^2[\alpha]$. Then from (4.10) we see that $s = mpq$ for some $p \in \overline{p}[\alpha]$ and some $q \in \overline{q}[\alpha]$ with $p + q = 1$. Hence $s \in m\overline{p}[\alpha]\overline{q}[\alpha]$ and $s \in (m\overline{p}\overline{q})[\alpha]$. The result follows. The next example shows that $\overline{\sigma}^2$ may not equal $m\overline{p}\overline{q}$.

Example 4.2.2

Let $\bar{p} = (0.4/0.6/0.8)$ and $\bar{q} = 1 - \bar{p} = (0.2/0.4/0.6)$. If p is in $\bar{p}[\alpha]$, then we can always choose $q = 1 - p \in \bar{q}[\alpha]$ so that $p + q = 1$. Then

$$\bar{\sigma}^2[\alpha] = \{mp(1 - p)|p \in \bar{p}[\alpha]\}, \tag{4.11}$$

from (4.10). Let $h(p) = mp(1 - p)$. We see that $h(p)$: (1) is increasing on $[0, 0.5]$; (2) has its maximum of 0.25 m at $p = 0.5$; and (3) is decreasing on $[0.5, 1]$. So, the evaluation of (4.11) depends if $p = 0.5$ belongs to the α-cut of $\bar{p}$. Let $\bar{p}[\alpha] = [p_1(\alpha), p_2(\alpha)] = [0.4 + 0.2\alpha, 0.8 - 0.2\alpha]$. So, $p = 0.5$ belongs to the α-cut of $\bar{p}$ only for $0 \le \alpha \le 0.5$. Then

$$\bar{\sigma}^2[\alpha] = [h(p_2(\alpha)), 0.25m], \tag{4.12}$$

for $0 \le \alpha \le 0.5$, and

$$\bar{\sigma}^2[\alpha] = [h(p_2(\alpha)), h(p_1(\alpha))], \tag{4.13}$$

for $0.5 \le \alpha \le 1$. We substitute in for $p_1(\alpha)$ and $p_2(\alpha)$ to finally obtain $\bar{\sigma}^2$ and its graph, for $m = 10$, using Maple [5] is in Fig. 4.1. Clearly $\bar{\sigma}^2$ does not equal $10\bar{p}\bar{q}$. For example $\bar{\sigma}^2[0] = [1.6, 2.5]$ but $(10\bar{p}\bar{q})[0] = [0.8, 4.8]$.

4.3 Fuzzy Poisson

Let X be a random variable having the Poisson probability mass function. If $P(x)$ stands for the probability that $X = x$, then

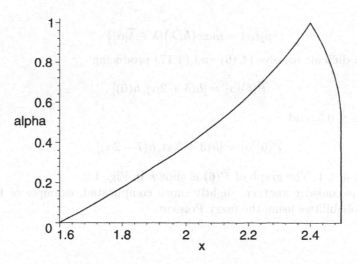

Fig. 4.1. Fuzzy Variance in Example 4.2.2

$$P(x) = \frac{\lambda^x \exp(-\lambda)}{x!} , \tag{4.14}$$

for $x = 0, 1, 2, 3, \ldots$, and parameter $\lambda > 0$. Now λ is not known precisely and must be estimated from some (crisp) data. See Sect. 3.4.1. Our estimator is a fuzzy number so substitute this fuzzy number $\overline{\lambda} > 0$ for λ to produce the fuzzy Poisson probability mass function. Set $\overline{P}(x)$ to be the fuzzy probability that $X = x$. Then we find α-cuts of this fuzzy number as

$$\overline{P}(x)[\alpha] = \left\{ \frac{\lambda^x \exp(-\lambda)}{x!} | \lambda \in \overline{\lambda}[\alpha] \right\} , \tag{4.15}$$

for all $\alpha \in [0, 1]$. The evaluation of (4.15) depends on the relation of x to $\overline{\lambda}[0]$. Let $h(\lambda) = \frac{\lambda^x \exp(-\lambda)}{x!}$ for fixed x and $\lambda > 0$. We see that $h(\lambda)$ is an increasing function of λ for $\lambda < x$, the maximum value of $h(\lambda)$ occurs at $\lambda = x$, and $h(\lambda)$ is a decreasing function of λ for $\lambda > x$. Let $\overline{\lambda}[\alpha] = [\lambda_1(\alpha), \lambda_2(\alpha)]$, for $0 \leq \alpha \leq 1$. Then we see that: (1) if $\lambda_2(0) < x$, then $\overline{P}(x)[\alpha] = [h(\lambda_1(\alpha)), h(\lambda_2(\alpha))]$; and (2) if $x < \lambda_1(0)$, then $\overline{P}(x)[\alpha] = [h(\lambda_2(\alpha)), h(\lambda_1(\alpha))]$. The other case, where $x \in \overline{\lambda}[0]$, is explored in the following example.

Example 4.3.1

Let $x = 6$ and $\overline{\lambda} = (3/5/7)$. We see that $x \in [3, 7] = \overline{\lambda}[0]$. We determine $\overline{\lambda}[\alpha] = [3 + 2\alpha, 7 - 2\alpha]$. Define $\overline{P}(6)[\alpha] = [p_1(\alpha), p_2(\alpha)]$. To determine the α-cuts of $\overline{P}(6)$ we need to solve (see Sect. 2.4.1)

$$p_1(\alpha) = min\{h(\lambda) | \lambda \in \overline{\lambda}[\alpha]\} , \tag{4.16}$$

and

$$p_2(\alpha) = max\{h(\lambda) | \lambda \in \overline{\lambda}[\alpha]\} . \tag{4.17}$$

It is not difficult to solve (4.16) and (4.17) producing

$$\overline{P}(6)[\alpha] = [h(3 + 2\alpha), h(6)] , \tag{4.18}$$

for $0 \leq \alpha \leq 0.5$, and

$$\overline{P}(6)[\alpha] = [h(3 + 2\alpha), h(7 - 2\alpha)] , \tag{4.19}$$

for $0.5 \leq \alpha \leq 1$. The graph of $\overline{P}(6)$ is shown in Fig. 4.2.

Let us consider another, slightly more complicated, example of finding fuzzy probabilities using the fuzzy Poisson.

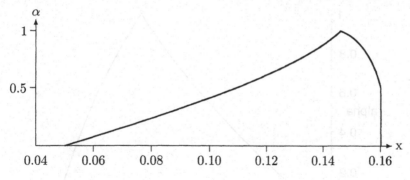

Fig. 4.2. Fuzzy Probability in Example 4.3.1

Example 4.3.2

Let $\overline{\lambda} = (8/9/10)$ and define $\overline{P}([3, \infty))$ to be the fuzzy probability that $X \geq 3$. Also let $\overline{P}([3, \infty))[\alpha] = [q_1(\alpha), q_2(\alpha)]$. Then

$$q_1(\alpha) = min\left\{1 - \sum_{x=0}^{2} h(\lambda) | \lambda \in \overline{\lambda}[\alpha]\right\}, \tag{4.20}$$

and

$$q_2(\alpha) = max\left\{1 - \sum_{x=0}^{2} h(\lambda) | \lambda \in \overline{\lambda}[\alpha]\right\}, \tag{4.21}$$

for all α. Let $k(\lambda) = 1 - [\sum_{x=0}^{2} h(\lambda)]$. Then $dk/d\lambda > 0$ for $\lambda > 0$ Hence, we may evaluate (4.20) and (4.21) and get

$$\overline{P}([3, \infty))[\alpha] = [k(\lambda_1(\alpha)), k(\lambda_2(\alpha))]. \tag{4.22}$$

This fuzzy probability is shown in Fig. 4.3. The graph in Fig. 4.3 was also constructed by Maple [5].

To finish this section we now compute the fuzzy mean and the fuzzy variance of the fuzzy Poisson probability mass function. Alpha-cuts of the fuzzy mean are

$$\overline{\mu}[\alpha] = \left\{\sum_{x=0}^{\infty} xh(\lambda) | \lambda \in \overline{\lambda}[\alpha]\right\}, \tag{4.23}$$

which reduces to, since the mean of the crisp Poisson is λ, the expression

$$\overline{\mu}[\alpha] = \{\lambda | \lambda \in \overline{\lambda}[\alpha]\}. \tag{4.24}$$

Hence, $\overline{\mu} = \overline{\lambda}$. So the fuzzy mean is just the fuzzification of the crisp mean. Let the fuzzy variance be $\overline{\sigma}^2$ and we obtain its α-cuts as

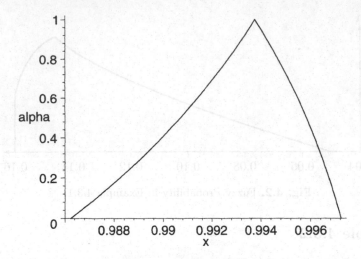

Fig. 4.3. Fuzzy Probability in Example 4.3.2

$$\overline{\sigma}^2[\alpha] = \left\{ \sum_{x=0}^{\infty} (x-\mu)^2 h(\lambda) | \lambda \in \overline{\lambda}[\alpha], \mu = \lambda \right\}, \tag{4.25}$$

which reduces to, since the variance of the crisp Poisson is also λ, the expression

$$\overline{\sigma}^2[\alpha] = \{\lambda | \lambda \in \overline{\lambda}[\alpha]\}. \tag{4.26}$$

It follows that $\overline{\sigma}^2 = \overline{\lambda}$ and the fuzzy variance is the fuzzification of the crisp variance.

4.4 Fuzzy Normal

The normal density $N(\mu, \sigma^2)$ has density function $f(x; \mu, \sigma^2)$, $x \in \mathbf{R}$, mean μ and variance σ^2. If the mean and variance are unknown we must estimate them from a random sample and we obtain fuzzy estimator $\overline{\mu}$ (Sect. 3.5.2) for μ and fuzzy estimator $\overline{\sigma}^2$ (Sect. 3.5.3) for σ^2. So consider the fuzzy normal $N(\overline{\mu}, \overline{\sigma}^2)$ for fuzzy numbers $\overline{\mu}$ and $\overline{\sigma}^2 > 0$. We wish to compute the fuzzy probability of obtaining a value in the interval $[c, d]$. We write this fuzzy probability as $\overline{P}[c, d]$. We may easily extend our results to $\overline{P}[E]$ for other subsets E of $\mathbf{R}$. For $\alpha \in [0, 1]$, $\mu \in \overline{\mu}[\alpha]$ and $\sigma^2 \in \overline{\sigma}^2[\alpha]$ let $z_1 = (c - \mu)/\sigma$ and $z_2 = (d - \mu)/\sigma$. Then

$$\overline{P}[c, d][\alpha] = \left\{ \int_{z_1}^{z_2} f(x; 0, 1) dx | \mu \in \overline{\mu}[\alpha], \sigma^2 \in \overline{\sigma}^2[\alpha] \right\}, \tag{4.27}$$

for $0 \leq \alpha \leq 1$. The above equation gets the α-cuts of $\overline{P}[c, d]$. Also, in the above equation $f(x; 0, 1)$ stands for the standard normal density with zero mean and unit variance. Let $\overline{P}[c,d][\alpha] = [p_1(\alpha), p_2(\alpha)]$. Then the minimum (maximum) of the expression on the right side of the above equation is $p_1(\alpha)$ $(p_2(\alpha))$. In general, it will be difficult to find these minimums (maximums) and one might consider using a genetic (evolutionary) algorithm, or some other numerical technique. However, as the following example shows, in some cases we can easily compute these α-cuts.

Example 4.4.1

Suppose $\overline{\mu} = (8/10/12)$, or the mean is approximately 10, and $\overline{\sigma}^2 = (4/5/6)$, or the variance is approximately five. Compute $\overline{P}[10,15]$. First it is easy to find the $\alpha = 1$ cut and we obtain $\overline{P}[10,15][1] = 0.4873$. Now we want the $\alpha = 0$ cut. Using the software package Maple [5] we graphed the function

$$g(x,y) = \int_{z_1}^{z_2} f(u; 0, 1) du , \tag{4.28}$$

for $z_1 = (10-x)/y$, $z_2 = (15-x)/y$, $8 \leq x \leq 12$, $4 \leq y^2 \leq 6$. Notice that the $\alpha = 0$ cut of $(8/10/12)$ is $[8,12]$, the range for $x = \mu$, and of $(4/5/6)$ is $[4,6]$ the range for $y^2 = \sigma^2$. The surface clearly shows: (1) a minimum of 0.1584 at $x = 8$ and $y = 2$; and (2) a maximum of 0.7745 at $x = 12$ and $y = 2$. Hence the $\alpha = 0$ cut of this fuzzy probability is $[0.1584, 0.7745]$. But from this graph we may also find other α-cuts. We see from the graph that $g(x,y)$ is an increasing function of: (1) x for y fixed at a value between 2 and $\sqrt{6}$; and (2) y for x fixed at 8. However, $g(x,y)$ is a decreasing function of y for $x = 12$. This means that for any α-cut: (1) we get the max at $y =$ its smallest value and $x =$ at its largest value; and (2) we have the min when $y =$ at is smallest and $x =$ its least value. Some α-cuts of $\overline{P}[10,15]$ are shown in Table 4.1 and Fig. 4.4 displays this fuzzy probability. The graph in Fig. 4.4 is only an approximation because we did not force the graph through all the points in Table 4.1.

Table 4.1. Alpha-cuts of the Fuzzy Probability in Example 4.4.1

α	$\overline{P}[10,15][\alpha]$
0	[0.1584,0.7745]
0.2	[0.2168,0.7340]
0.4	[0.2821,0,6813]
0.6	[0.3512,0.6203]
0.8	[0.4207,0.5545]
1.0	[0.4873,0.4873]

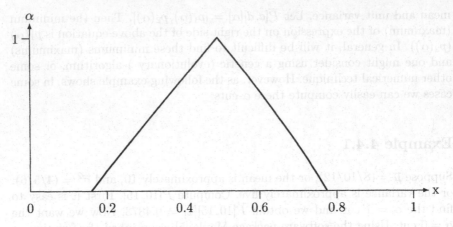

Fig. 4.4. Approximate Fuzzy Probability in Example 4.4.1

We now show that the fuzzy mean of $N(\overline{\mu}, \overline{\sigma}^2)$ is $\overline{\mu}$ and the fuzzy variance is $\overline{\sigma}^2$, respectively, the fuzzification of the crisp mean and variance. Let the fuzzy mean be $\overline{M}$. Then its α-cuts are

$$\overline{M}[\alpha] = \left\{ \int_{-\infty}^{\infty} x f(x; \mu, \sigma^2) dx | \mu \in \overline{\mu}[\alpha], \sigma^2 \in \overline{\sigma}^2[\alpha] \right\} . \tag{4.29}$$

But the integral in the above equation equals μ for any $\mu \in \overline{\mu}[\alpha]$ and any $\sigma^2 \in \overline{\sigma}^2[\alpha]$. Hence $\overline{M} = \overline{\mu}$. Let the fuzzy variance be $\overline{V}$. Then its α-cuts are

$$\overline{V}[\alpha] = \left\{ \int_{-\infty}^{\infty} (x - \mu)^2 f(x, \mu, \sigma^2) dx | \mu \in \overline{\mu}[\alpha], \sigma^2 \in \overline{\sigma}^2[\alpha] \right\} . \tag{4.30}$$

We see that the integral in the above equation equals σ^2 for all $\mu \in \overline{\mu}[\alpha]$ and all $\sigma^2 \in \overline{\sigma}^2[\alpha]$. Therefore, $\overline{V} = \overline{\sigma}^2$.

4.5 Fuzzy Exponential

The exponential $E(\lambda)$ has density $f(x; \lambda) = \lambda \exp(-\lambda x)$ for $x \geq 0$ and $f(x; \lambda) = 0$ otherwise, where $\lambda > 0$. The mean and variance of $E(\lambda)$ is $1/\lambda$ and $1/\lambda^2$, respectively. If the parameter λ is unknown we estimate it from some random sample producing fuzzy estimator $\overline{\lambda}$ (Sect. 3.4.2) for λ. Now consider $E(\overline{\lambda})$ for fuzzy number $\overline{\lambda} > 0$. Let us find the fuzzy probability of obtaining a value in the interval $[c, d]$, $c > 0$. Denote this probability as $\overline{P}[c, d]$. One may generalize to $\overline{P}[E]$ for other subsets E of $\mathbf{R}$. We compute

$$\overline{P}[c, d][\alpha] = \left\{ \int_{c}^{d} \lambda \exp(-\lambda x) dx | \lambda \in \overline{\lambda}[\alpha] \right\} , \tag{4.31}$$

for all α. Let $\overline{P}[c,d][\alpha] = [p_1(\alpha), p_2(\alpha)]$, then

$$p_1(\alpha) = min\left\{\int_c^d \lambda \exp(-\lambda x)dx | \lambda \in \overline{\lambda}[\alpha]\right\}, \qquad (4.32)$$

and

$$p_2(\alpha) = max\left\{\int_c^d \lambda \exp(-\lambda x)dx | \lambda \in \overline{\lambda}[\alpha]\right\}, \qquad (4.33)$$

for $0 \le \alpha \le 1$. Let

$$h(\lambda) = \exp(-c\lambda) - \exp(-d\lambda) = \int_c^d \lambda \exp(-\lambda x)dx, \qquad (4.34)$$

and we see that h: (1) is an increasing function of λ for $0 < \lambda < \lambda^*$; and (2) is a decreasing function of λ for $\lambda^* < \lambda$. We find that $\lambda^* = -[ln(c/d)]/(d-c)$. Assume that $\overline{\lambda} > \lambda^*$. So we can now easily find $\overline{P}[c,d]$. Let $\overline{\lambda}[\alpha] = [\lambda_1(\alpha), \lambda_2(\alpha)]$. Then

$$p_1(\alpha) = h(\lambda_2(\alpha)), \qquad (4.35)$$

and

$$p_2(\alpha) = h(\lambda_1(\alpha)). \qquad (4.36)$$

We give a picture of this fuzzy probability, using Maple [5], in Fig. 4.5 when: (1) $c = 1$ and $d = 4$; and (2) $\overline{\lambda} = (1/3/5)$.

Next we find the fuzzy mean and fuzzy variance of $E(\overline{\lambda})$. If $\overline{\mu}$ denotes the mean, we find its α-cuts as

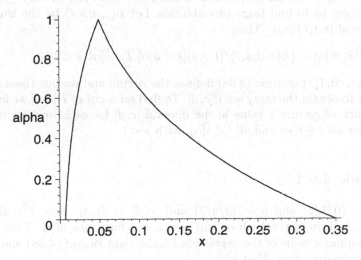

Fig. 4.5. Fuzzy Probability for the Fuzzy Exponential

$$\overline{\mu}[\alpha] = \left\{ \int_0^\infty x\lambda \exp(-\lambda x)dx | \lambda \in \overline{\lambda}[\alpha] \right\}, \qquad (4.37)$$

for all α. However, each integral in the above equation equals $1/\lambda$. Hence $\overline{\mu} = 1/\overline{\lambda}$. If $\overline{\sigma}^2$ is the fuzzy variance, then we write down an equation to find its α-cuts we obtain $\overline{\sigma}^2 = 1/\overline{\lambda}^2$. The fuzzy mean (variance) is the fuzzification of the crisp mean (variance).

4.6 Fuzzy Uniform

The uniform density $U(a,b)$, $a < b$, has density function $y = f(x;a,b) = 1/(b-a)$ for $a < x < b$ and $f(x;a,b) = 0$ otherwise. Now consider $U(\overline{a},\overline{b})$ for fuzzy numbers $\overline{a}$ and $\overline{b}$. In this section we will model $\overline{a}$ and $\overline{b}$ as triangular fuzzy numbers. Notice that in Sect. 3.5.5 $\overline{a}$ and $\overline{b}$ were half of a triangular shaped fuzzy number as shown in Fig. 3.9 and the left side of $\overline{a}$ got cut off in Fig. 3.8. The fuzzy sets in Figs. 3.8 and 3.9 are what we should use in the rest of the book but for now let us have triangular fuzzy numbers for $\overline{a}$ and $\overline{b}$. Assume that $\overline{a}[1] = a$ and $\overline{b}[1] = b$ so that $\overline{a}$ ($\overline{b}$) represents the uncertainty in a (b). Now using the fuzzy uniform density we wish to compute the fuzzy probability of obtaining a value in the interval $[c,d]$. Denote this fuzzy probability as $\overline{P}[c,d]$. We can easily generalize to $\overline{P}[E]$ for more general subsets E.

There is uncertainty in the end points of the uniform density but there is no uncertainty in the fact that we have a uniform density. What this means is that given any $s \in \overline{a}[\alpha]$ and $t \in \overline{b}[\alpha]$, $s < t$, we have a $U(s,t)$, or $f(x;s,t) = 1/(t-s)$ on $[s,t]$ and it equals zero otherwise, for all $0 \le \alpha \le 1$. This enables us to find fuzzy probabilities. Let $L(c,d;s,t)$ be the length of the interval $[s,t] \cap [c,d]$. Then

$$\overline{P}[c,d][\alpha] = \{L(c,d;s,t)/(t-s)|s \in \overline{a}[\alpha], t \in \overline{b}[\alpha], s < t\}, \qquad (4.38)$$

for all $\alpha \in [0,1]$. Equation (4.38) defines the α-cuts and we put these α-cuts together to obtain the fuzzy set $\overline{P}[c,d]$. To find an α-cut of $\overline{P}[c,d]$ we find the probability of getting a value in the interval $[c,d]$ for each uniform density $U(s,t)$ for all $s \in \overline{a}[\alpha]$ and all $t \in \overline{b}[\alpha]$, with $s < t$.

Example 4.6.1

Let $\overline{a} = (0/1/2)$ and $\overline{b} = (3/4/5)$ and $[c,d] = [1,4]$. Now $\overline{P}[c,d][\alpha] = [p_1(\alpha), p_2(\alpha)]$ an interval whose end points are functions of α. Then $p_1(\alpha)$ is the minimum value of the expression on the right side of (4.38) and $p_2(\alpha)$ is the maximum value. That is

$$p_1(\alpha) = min\{L(1,4;s,t)/(t-s)|s \in \overline{a}[\alpha], t \in \overline{b}[\alpha]\}, \qquad (4.39)$$

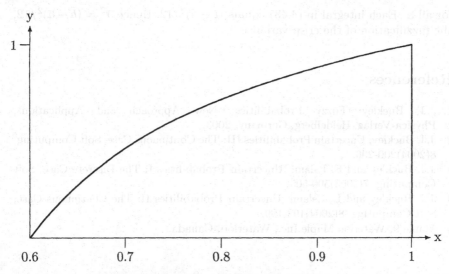

Fig. 4.6. Fuzzy Probability in Example 4.6.1

and

$$p_2(\alpha) = max\{L(1,4;s,t)/(t-s)|s \in \overline{a}[\alpha], t \in \overline{b}[\alpha]\} . \qquad (4.40)$$

It is easily seen that $p_2(\alpha) = 1$ all α in this example. To find the minimum we must consider four cases. First $\overline{a}[\alpha] = [\alpha, 2-\alpha]$ and $\overline{b}[\alpha] = [3+\alpha, 5-\alpha]$. Then the cases are: (1) $\alpha \le s \le 1, 3+\alpha \le t \le 4$; (2) $\alpha \le s \le 1, 4 \le t \le 5-\alpha$; (3) $1 \le s \le 2-\alpha, 3+\alpha \le t \le 4$; and (4) $1 \le s \le 2-\alpha, 4 \le t \le 5-\alpha$. Studying all four cases we obtain the minimum equal to $3/(5-2\alpha)$. Hence the α-cuts of $\overline{P}[1,4]$ are $[3/(5-2\alpha), 1]$ and the graph of this fuzzy number is in Fig. 4.6.

Next we want to find the mean and variance of $U(\overline{a}, \overline{b})$. Let the mean be $\overline{\mu}$ and we find its α-cuts as follows

$$\overline{\mu}[\alpha] = \left\{ \int_s^t (x/(t-s))dx | s \in \overline{a}[\alpha], t \in \overline{b}[\alpha], s < t \right\} , \qquad (4.41)$$

for all α. But each integral in (4.41) equals $(s+t)/2$. Hence, assuming $\overline{a}[0] = [s_1, s_2]$, $\overline{b}[0] = [t_1, t_2]$ and $s_2 < t_1$,

$$\overline{\mu} = (\overline{a} + \overline{b})/2 . \qquad (4.42)$$

So, $\overline{\mu}$ is the fuzzification of the crisp mean $(a+b)/2$. If the variance of $U(\overline{a}, \overline{b})$ is $\overline{\sigma}^2$, then its α-cuts are

$$\overline{\sigma}^2[\alpha] = \left\{ \int_s^t [(x-\mu)^2/(t-s)]dx | s \in \overline{a}[\alpha], t \in \overline{b}[\alpha], \mu = (s+t)/2, s < t \right\} , \qquad (4.43)$$

for all α. Each integral in (4.43) equals $(t-s)^2/12$. Hence $\overline{\sigma}^2 = (\overline{b} - \overline{a})^2/12$, the fuzzification of the crisp variance.

References

1. J.J. Buckley: Fuzzy Probabilities: New Approach and Applications, Physica-Verlag, Heidelberg, Germany, 2003.
2. J.J. Buckley: Uncertain Probabilities III: The Continuous Case, Soft Computing, 8(2004)200-206.
3. J.J. Buckley and E. Eslami: Uncertain Probabilities I: The Discrete Case, Soft Computing, 7(2003)500-505.
4. J.J. Buckley and E. Eslami: Uncertain Probabilities II: The Continuous Case, Soft Computing, 8(2004)193-199.
5. Maple 9, Waterloo Maple Inc., Waterloo, Canada.

5 Fuzzy Systems Theory

5.1 Introduction

In this chapter we consider systems whose performance depends on probability distributions and some of the parameters in these probability distributions are to be estimated. We argue that all such systems become fuzzy systems whose performance will depend on fuzzy probability distributions and whose measures of performance can be described by fuzzy numbers. This chapter is based on [1].

We will consider one example to illustrate our argument. This example, of a queuing network, is shown in Fig. 5.1. We first describe the crisp (non-fuzzy) system in Fig. 5.1 and then, in the next section, we discuss why, and how, it becomes fuzzy. A brief discussion on how simulation may be used to estimate fuzzy measures of system performance is in Sect. 5.3.

Items arrive at the arrival node according to the exponential distribution with mean 1.5 minutes. Time within the system will be measured in minutes. These items go immediately to queue number one (Q1 in Fig. 5.1). This queue is assumed to have unlimited capacity so that we never loose an arrival due to the system being full. All queues can have any queue discipline (first come, first served; priority, etc.). The first server in the system is labelled S1. An item from Q1 enters S1 when S1 is available. All service times are assumed to be normally distributed. Let $N(\mu, \sigma^2)$ be the normal distribution with mean μ and variance σ^2. So assume that $N(\mu_1, \sigma_1^2)$ describes the service time for S1. For all $N(\mu, \sigma^2)$ used for service time we assume that $\mu - 5\sigma^2 > 0$ so that we will not be able to randomly generate negative service times using the normal distribution.

After service in S1 the item enters the queue Q2 for the next servers S2. All other queues (Qi, i = 2,3,4,5) in Fig. 5.1 have finite capacity. If Q2 is full, then an item in S1 can not leave S1, even though its service there is completed, until there is space for it in Q2. S2 consists of two parallel and identical servers S21 and S22 both with service times $N(\mu_2, \sigma_2^2)$. After completing service in S21 or S22 the item goes to queue Q3 for server S3 whose service time is $N(\mu_3, \sigma_3^2)$. Then on to Q4 and three parallel and identical servers S41, S42, S43 each with service time $N(\mu_4, \sigma_4^2)$. Next is Q5 for the inspection service stations S51 and S52. The inspector at S51 is more experienced and faster than the inspector at S52 so: (1) Q5 always feeds S51 first if it is available; and (2) Q5

James J. Buckley: *Simulating Fuzzy Systems*, StudFuzz **171**, 49–52 (2005)
www.springerlink.com   © Springer-Verlag Berlin Heidelberg 2005

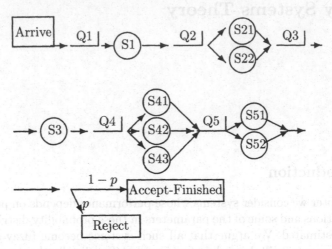

Fig. 5.1. Queuing System

sends an item to S52 only when S51 is busy and S52 is available. Their service times are $N(\mu_{5i}, \sigma_5^2)$, $i = 1, 2$ with $\mu_{51} < \mu_{52}$. Finally, on the average $100p\%$ of the items are rejected since they do not meet quality constraints and the other $100(1 - p)\%$ are acceptable and these become the finished product.

We may easily simulate this system (see Chap. 6). Some of the statistics (performance measures) we are interested in are: (1) $R =$ the average response time, or the average time an acceptable (not rejected) item spends in the system; (2) $X =$ throughput, or the expected number of acceptable items produced per unit time; (3) $U_i =$ various average utilizations of the servers; and (4) $N =$ the expected number of units in the system. The simulation can produce these, and other, system statistics. Then we may want to "optimize" the system. For example, we may want to configure the system differently to reduce R, increase X, etc. We consider this system again in Chap. 9.

5.2 Fuzzy System

The performance of the system described above depends on probability distributions: (1) the exponential with mean 1.5 minutes; (2) the normal distribution for service times; and (3) the Bernoulli distribution with probability p of a defective. We assume that some of the parameters in these distributions have been estimated from data. From this data we construct a set of confidence intervals for the unknown parameter, and then place these confidence intervals one on top of another to obtain fuzzy number estimators for these parameters (see Chap. 3). This produces fuzzy probability distributions. This procedure has been discussed before in Chaps. 3 and 4. With fuzzy probability distributions we will then compute fuzzy numbers for system performance and we have a fuzzy system.

In crisp systems that depend on probability distributions one usually obtains point estimates of the parameters in these distributions to be used to calculate system performance. However, point estimates show no uncertainty. There can be a lot of uncertainty in the model due to these parameters not being known precisely.

We suggest adding uncertainty through using fuzzy estimators for the unknown parameters. The fuzzy estimators are constructed from a set of confidence intervals. The fuzzy estimators produce fuzzy probability distributions which then make the system into a fuzzy system. Hence, many crisp systems become fuzzy systems through this procedure. System performance measures will now be described by fuzzy numbers which contain the uncertainties from the estimation of the parameters in the probability distributions.

Computation of system performance measures becomes much more difficult in the fuzzy system. Assuming the crisp system is sufficiently complicated we will need to simulate it to estimate these measures. Chapter 7 discusses why simulation can produce correct estimates for the fuzzy numbers for system performance. What simulation language we will use to estimate these fuzzy numbers is presented in the next chapter. Also, estimating the fuzzy numbers for system performance involves simulation optimization the topic of Chap. 8.

5.3 Computing Fuzzy Measures of Performance

Now we need to compute $\overline{R}$, $\overline{X}$, $\overline{N}$ and the $\overline{U}_i$. They will be determined by their α-cuts. We will consider calculating $\overline{R}[\alpha]$, $0 \leq \alpha \leq 1$, since the others are computed in a similar manner. We first discuss approximating the alpha-cuts using simulation and then why simulation will produce the correct result.

Let **SIM** denote some simulation software (Chap. 6) that can be used to simulate the crisp system in Fig. 5.1. **SIM** sends items through the system until we get N accepted (not rejected) items. N could be 1000, or $500,000$, etc. From this simulation we obtain a value for R. Actually, the simulation makes a distribution of values for time in the system and the R we want is the expected value (mean value) of this distribution. So now assume that the output from **SIM** will be the mean values we want. Input to **SIM** will be λ (for the exponential), the means μ_i and variances σ_i^2 of the normal distributions for service time and p the probability of a defective. We summarize all this as

$$R = \mathbf{SIM}(\lambda, \mu_1, \sigma_1^2, \ldots, \mu_{52}, \sigma_5^2, p) . \tag{5.1}$$

Now go to the fuzzy system. We obtain fuzzy number estimators $\overline{\lambda}$, $\overline{\mu}_1, \ldots, \overline{p}$. Then

$$\overline{R}[\alpha] = \{R | R = \mathbf{SIM}(\lambda, \mu_1, \ldots, \sigma_5^2, p) , \mathcal{S}\} , \tag{5.2}$$

where
$$S = \lambda \in \overline{\lambda}[\alpha], \mu_1 \in \overline{\mu}_1[\alpha], \ldots, \sigma_5^2 \in \overline{\sigma}_5^2[\alpha], p \in \overline{p}[\alpha] \,, \tag{5.3}$$

for $0 \leq \alpha \leq 1$. Alpha-cuts of $\overline{R}$ will be intervals so let $\overline{R}[\alpha] = [r_1(\alpha), r_2(\alpha)]$. Then

$$r_1(\alpha) = min\{R | R = \mathbf{SIM}(\lambda, \mu_1, \ldots, \sigma_5^2, p), S\} \,, \tag{5.4}$$

and

$$r_2(\alpha) = max\{R | R = \mathbf{SIM}(\lambda, \mu_1, \ldots, \sigma_5^2, p), S\} \,. \tag{5.5}$$

So we solve (5.4) and (5.5) for selected values of α and we have estimated the fuzzy number for the expected time an item spends in the system. Also, $\overline{R}[0]$ is like a 99% confidence interval for this time. We compute the other fuzzy numbers $(\overline{X}, \overline{N}, \overline{U}_i)$ the same way.

How are we to solve (5.4) and (5.5)? We suggest a simulation optimization method discussed in Chap. 8. The search space is given by S. As we shall see using the end points of the intervals for the α-cuts of $\overline{\lambda}, \overline{\mu}_1, \ldots, \overline{p}$ will solve many of these problems.

Now assume we have a function F to compute R from the values of the parameters. That is

$$R = F(\lambda, \mu_1, \ldots, \sigma_5^2, p) \,. \tag{5.6}$$

We do not have such a function but let us assume that it exists and is continuous. Then, using the extension principle (Sect. 2.4.1), we have

$$\overline{R} = F(\overline{\lambda}, \overline{\mu}_1, \ldots, \overline{p}) \,. \tag{5.7}$$

Then we can argue, see Chap. 7, that simulation can be used to approximate α-cuts of $\overline{R}$ obtained from the extension principle. The computation of R in (5.6) is what we call a "one-step" method since we use only one function. In Chap. 7 we show that simulation can be used to get correct results, approximating $R[\alpha]$, when R is determined by the "one-step" procedure. In fact, we must employ simulation to approximate $\overline{R}$, $\overline{X}$, $\overline{N}$ and the $\overline{U}_i$ when such a one-step function exists but is unknown. In most applications we will not know the one-step function.

Reference

1. J.J. Buckley, Fuzzy Systems, Soft Computing. To appear.

6 Simulation

Now we come to the point were we need to select simulation software to do all the crisp simulations staring in Chap. 7. The author is not an expert in simulation and does not know about all the products that are available. We will do a "search" for simulation products.

In choosing a simulation package we will have two main constraints: (1) it must be inexpensive, hopefully at most 100 US dollars; and (2) it must be easy to use. Also, it has to run on a desktop computer and collect the statistics we are interested in (throughput, response time,...). To start the search put "simulation" into your web search engine and we obtained too many pages to study. However, we are not interested in all types of simulation, only "discrete event" simulation. In discrete event simulation, events like an item leaving a server or a customer arriving at a queue, can occur only at certain times. The state of the model changes only at discrete event times. The opposite of discrete event simulation would be continuous simulation which models systems evolving through continuous time usually governed by systems of differential equations. So we put discrete event simulation into our web search engine and came up with around 20 pages to study. Looking at the web sites on these pages was interesting but, as we found out later, not all the discrete event simulation products that we would like to see were listed. The best result of this search was that the author found a recent survey of discrete event simulation vendors [6]. This survey lists 48 products with information about the software, including typical applications, primary market, animation, system requirements, price, and if an optimization program is included, together with a web site. However, such lists can become quickly outdated because company/product names can change due to mergers, acquisitions, etc.

Most simulation software comes as two types: (1) a "scaled down" version usually called the "student version"; and (2) the full version usually called the "standard" or "professional" version. The professional software is too expensive, usually in the thousands of dollars, so we are interested in using the student software. The student version has some limitations, usually at most 50 (or maybe as high as 200) stations (service, queue,...) can be used. You may be able to download the student (and professional) software and use it for free for a certain time period. Also, you can sometimes get the student

James J. Buckley: *Simulating Fuzzy Systems*, StudFuzz **171**, 53–55 (2005)
www.springerlink.com

version free if it comes with a book, like a user's manual, that you purchase. Too often, if you have to buy it the student software is priced too high for our budget (max $100).

We also want the simulation software self contained and ready to use so we do not need to write code (C++, Java, ...) to run a simulation. Obviously, it must have a good user's manual. We first decided to use the "click-drag-drop" simulation software. The "click-drag-drop" method makes it very easy to construct a network. Each item in your system, service station, queue, branch,.. has an icon in the tool bar which you click on, then drag it to where you want it in the network and then drop it. Separate windows open for you to input information about the icon (like service time, arrival rates, ...). We found this the easiest method to build the systems we want to simulate.

The book [3] contains a chapter devoted to selecting discrete event simulation software. Most of the points addressed here pertain to corporation's use of simulation. Let us briefly look at some of the major topics. They are

1. Functionality: does it exhibit required functions to satisfy your particular needs? Under this heading is that it should provide for steady-state performance analysis. When the system first starts up it exhibits transient (or warm-up) behavior and then settles into steady-state. The statistics collected should not be biased by the warm-up time.
2. Usability: how much effort is needed to learn and use the software?
3. Reliability: are there any "bugs"?
4. Maintainability: how much effort is needed to upgrade. What is the documentation? What are the hardware requirements?
5. Scalability: the ability of the product to function at multiple levels under various skill levels.
6. Vendor quality: what do you know about the company?
7. Vendor services: what training, technical support and consulting services are available?
8. Cost: What are the total costs (buy it, maintenance, training, run times, etc.) associated with this software over n years?

For this book we initially decided on SIMPROCESS [2,3,7] because: (1) it has the "click-drag-drop" method of modelling; (2) it has a good user's manual that you can download from their website [5] for free and you can get it, and learn to use the product, before you get the software; (3) you can download both the student's version and the professional version to use for free for a limited time (varies from 3 to 8 weeks); (4) it runs on a desktop computer and obtains all the summary statistics we want; and (5) technical support appears very good which was learned through numerous emails.

However, the free use time for the student's version will probably expire before this book is completed. That would be a disaster. So we switched to using GPSS. We obtained the GPSS manual [1] which contained the CD-ROM with the student's version of GPSS/H for free. This simulation language

(GPSS) has been around for more than 20 years and an excellent introduction can be found in the "big red book" [4]. More information about GPSS can be obtained at [8]. However, the student version has certain limitations: (1) at most 125 blocks; (2) at most 250 statements; and (3) 32,720 bytes of COMMON storage. In the simulations in this book we never had a problem with the constraints (1) and (2) above. However, we did run out of memory (COMMON).

The machine used for all simulations had 256 MB RAM. We would run out of COMMON (memory) for the following reasons. An item passing through the system during a simulation will be called a transaction. Each transaction has a file attached to it describing where it has been and where it is now at each tick of the clock in a discrete event simulation. Consider a system that has lots of queues with unlimited capacity, many arrivals of new transactions and service stations with slow service time. As time goes by the queues fill up with 100, 200, .. transactions with their larger and larger files. You will run out of COMMON even with 512 MB RAM. We would decrease the simulation time, hoping to complete the simulation before we ran out of COMMON. However, a too short simulation time is to be avoided. The simulation run time should be sufficiently long to get the system well into "steady-state" [1] so that the effect of the initial conditions become insignificant. In one, or two, cases we may have reduced run time too much in order to finish the simulation.

All simulations were done using GPSS and some of the programs are listed in Chap. 28. In Chap. 28 we do not present a tutorial on GPSS, we only list some of the programs with minimal comments.

References

1. J. Banks, J.S. Carson and J.N. Sy: Getting Started with GPSS/H, Second Edition, Wolverine Software Corp. Annandale, VA, 1995.
2. Getting Started SIMPROCESS, CACI Products Company, San Diego, CA, 2004.
3. H.J. Harrington and K. Tumay, Simulation Modelling Methods, McGraw-Hill, New York, 2000.
4. Thomas J. Schriber: Simulation Using GPSS, John Wiley and Sons, New York, 1974.
5. www.simprocess.com
6. J.J. Swain, Simulation Software Survey, OR/MS Today, August 2003, 46-57.
7. User's Manual SIMPROCESS, release 4, CACI Products Company, San Diego, CA, 2004.
8. www.wolverinesoftware.com

7 Queuing I: One-Step Calculations

7.1 Introduction

In this chapter we show situations where simulation can produce the same results as fuzzy calculations which employ the extension principle. We argue this result using an example problem in the next section. Then, in the third section, we show cases where simulation may not give the same answers as fuzzy computations. The last section presents a brief overlook of the rest of the book. In this chapter we assume the reader is familiar with the basic concepts of elementary queuing theory as may be found in [8].

7.2 One-Step Calculations

Let us consider a simple queuing system shown in Fig. 7.1. Customers arriving according to an exponential inter-arrival time at rate λ per time unit are placed into a queue if all servers are busy, or go directly into service when the queue is empty and a server is available. The calling source, where the customers come from, is very large and can be assumed to be infinite. The queue discipline is "first come, first served". The system has two parallel and identical servers ($c = 2$, labelled "S" in Fig. 7.1) with exponential inter-service time at rate μ per time unit. System capacity (queue plus servers) is $M = 4$ customers. We are interested in computing U = server utilization = the average fraction of time that at least one of the servers (assuming that there are multiple parallel servers) is busy, N = the expected number of customers in the system, X = average server throughput = the expected number of customers leaving the system per unit time, R = average response time = the expected time a customer spends in the system, assuming they entered the system when it was not at capacity and LC = the number of lost customers per unit time due to finite system capacity.

We may look up formulas for U, N, X, R and LC in terms of λ and μ, in any book that discusses elementary queuing theory. However, we first need expressions for the steady-state probabilities. Let p_i be the probability (steady-state) that there are i customers in the system, $0 \leq i \leq 4$. Then if $\rho = \lambda/\mu$

James J. Buckley: *Simulating Fuzzy Systems*, StudFuzz **171**, 57–67 (2005)
www.springerlink.com

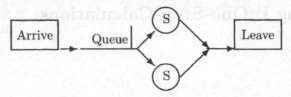

Fig. 7.1. Elementary Queuing System

$$p_0 = \left[1 + \rho + \frac{\rho^2(1 - \{\rho^3/8\})}{2(1 - \rho/2)}\right]^{-1} , \tag{7.1}$$

and $p_1 = \rho p_0$, $p_2 = (\rho^2/2)p_0$, $p_3 = (\rho^3/4)p_0$ and $p_4 = (\rho^4/8)p_0$. From these expressions for the p_i in terms of λ and μ we may obtain formulas for the U, N, X, R and LC involving only λ and μ.

$$U = f_u(\lambda, \mu) = p_1 + p_2 + p_3 + p_4 = 1 - p_0 . \tag{7.2}$$

Also

$$N = f_n(\lambda, \mu) = p_1 + 2p_2 + 3p_3 + 4p_4 , \tag{7.3}$$

and [3]

$$X = f_x(\lambda, \mu) = \mu p_1 + 2\mu(p_2 + p_3 + p_4) . \tag{7.4}$$

Next, $R = N/X$ so

$$R = f_r(\lambda, \mu) = \frac{p_1 + 2p_2 + 3p_3 + 4p_4}{\mu p_1 + 2\mu(p_2 + p_3 + p_4)} . \tag{7.5}$$

Finally,

$$LC = f_{lc}(\lambda, \mu) = \lambda p_4 . \tag{7.6}$$

The fact that we have such functions as f_u, f_n, f_x, f_r and f_{lc} is important because we can then compute U, N, X, R and LC in one-step. The "One-Step Calculations" in the title to this chapter means that we have such functions to compute what we want in "one-step".

7.2.1 Fuzzy Calculations

Now we must estimate both λ and μ from some crisp data so they both become fuzzy numbers $\overline{\lambda}$ and $\overline{\mu}$. See Chap. 3, Sect. 3.4. For this discussion assume that they are both triangular fuzzy numbers $\overline{\lambda} = (3/4/5)$ and $\overline{\mu} = (5/6/7)$. Since λ and μ are fuzzy U, N, X, R and LC will all become fuzzy numbers. We find $\overline{U}$, $\overline{N}$, $\overline{X}$, $\overline{R}$ and $\overline{LC}$ from the extension principle (Sect. 2.4.1 of Chap. 2). Then

$$\overline{U} = f_u(\overline{\lambda}, \overline{\mu}) , \tag{7.7}$$

$$\overline{N} = f_n(\overline{\lambda}, \overline{\mu}) , \tag{7.8}$$

$$\overline{X} = f_x(\overline{\lambda}, \overline{\mu}) , \tag{7.9}$$

$$\overline{R} = f_r(\overline{\lambda}, \overline{\mu}) , \tag{7.10}$$

and

$$\overline{LC} = f_{lc}(\overline{\lambda}, \overline{\mu}) . \tag{7.11}$$

There are some general constraints attached to computing the items in (7.7)–(7.11). They are: (1) $0 \leq \overline{U} \leq 1$; (2) $0 \leq \overline{N} \leq 4 =$ system capacity; and (3) $0 \leq \overline{X}, \overline{R}, \overline{LC}$. Given that $\lambda \in [3,5]$ and $\mu \in [5,7]$ other constraints are (1) $\overline{X} \leq 5 =$ maximum arrival rate; (2) $\overline{LC} \leq 5$; and (3) $\overline{R} \leq 3(1/5) = 0.6$. The last constraint occurs because the longest a customer can be in the system is three service completions at minimum service rate (maximum service time).

For now let us concentrate only on N and X. Let $\overline{N}[\alpha] = [n_1(\alpha), n_2(\alpha)]$ and $\overline{X}[\alpha] = [x_1(\alpha), x_2(\alpha)]$, for $0 \leq \alpha \leq 1$. Since f_n and f_x are continuous functions, we may compute the alpha-cuts of $\overline{N}$ and $\overline{X}$ by solving the following optimization problems (see Sect. 2.4.1)

$$n_1(\alpha) = min\{f_n(\lambda, \mu) | \lambda \in \overline{\lambda}[\alpha], \mu \in \overline{\mu}[\alpha]\} , \tag{7.12}$$

$$n_2(\alpha) = max\{f_n(\lambda, \mu) | \lambda \in \overline{\lambda}[\alpha], \mu \in \overline{\mu}[\alpha]\} , \tag{7.13}$$

$$x_1(\alpha) = min\{f_x(\lambda, \mu) | \lambda \in \overline{\lambda}[\alpha], \mu \in \overline{\mu}[\alpha]\} , \tag{7.14}$$

$$x_2(\alpha) = max\{f_x(\lambda, \mu) | \lambda \in \overline{\lambda}[\alpha], \mu \in \overline{\mu}[\alpha]\} , \tag{7.15}$$

for $0 \leq \alpha \leq 1$.

Since f_n is continuous we know there is a $\lambda_l \in \overline{\lambda}[\alpha]$ and a $\mu_l \in \overline{\mu}[\alpha]$ so that

$$n_1(\alpha) = f_n(\lambda_l, \mu_l) . \tag{7.16}$$

That is, the minimum is taken on for some value of λ and μ. These optimization problems are easily solved using SOLVER in EXCEL. SOLVER is an optimization package which is an add-in to Microsoft EXCEL. It is free and if your EXCEL does not have it, then contact Microsoft or [7]. See [6] for applications of SOLVER. Also, there is a $\lambda_u \in \overline{\lambda}[\alpha]$ and a $\mu_u \in \overline{\mu}[\alpha]$ so that

$$n_2(\alpha) = f_n(\lambda_u, \mu_u) . \tag{7.17}$$

Similarly, we can find values of $\lambda \in \overline{\lambda}[\alpha]$ and $\mu \in \overline{\mu}[\lambda]$ for $x_1(\alpha)$ and $x_2(\alpha)$. All of this development pertains also to $\overline{U}, \overline{R}$ and $\overline{LC}$.

Example 7.2.1.1

Let us now solve these optimization problems (see (7.12)–(7.15)) to get alpha-cuts of $\overline{U}, \overline{N}, \overline{X}, \overline{R}$ and $\overline{LC}$. We need these results to compare to the simulation studies in the next subsection. Table 7.1 shows the $\alpha = 0$ cut and the

Table 7.1. One-Step Alpha Cuts of $\overline{U}$, $\overline{N}$, $\overline{X}$, $\overline{R}$, $\overline{LC}$ ($c = 2$, $M = 4$)

	Alpha = Zero Cut	Alpha = One Cut
$\overline{U}$	[0.3525, 0.6522]	0.4969
$\overline{N}$	[0.4456, 1.1304]	0.7205
$\overline{X}$	[2.9737, 4.9223]	3.9504
$\overline{R}$	[0.1489, 0.2364]	0.1824
$\overline{LC}$	[0.0082, 0.2174]	0.0497

$\alpha = 1$ cut. To use a spread sheet (Excel) to solve an optimization problem can be confusing, so let us go through some of the details on finding the $\alpha = 0$ cut.

Excel is a spread sheet whose columns are labelled A,B,C,... and rows are labelled 1,2,3,... So "cell" B4 means the cell in the fourth row and B column. When we say $H2 = \mathcal{K}$ we mean put into cell H2 the formula/expression/ data $\mathcal{K}$.

1. $A1 = \lambda = 4$, $A2 = \mu = 6$ (the initial values of λ and μ)
2. $B1 = (8 * A2^4) * (2 * A2 - A1)$ (the numerator of p_0)
3. $B2 = 16 * A2^5 + 8 * A1 * A2^4 - A1^5$ (the denominator of p_0)
4. $B3 = B1/B2$ (is p_0)
5. $B4 = A1/A2$ (equals $\rho = \lambda/\mu$)
6. $C1 = B4 * B3$ (computes p_1)
7. $C2 = (B4^2/2) * B3$ (computes p_2)
8. $C3 = (B4^3/4) * B3$ (computes p_3)
9. $C4 = (B4^4/8) * B3$ (computes p_4)
10. $D1 = U = C1 + C2 + C3 + C4$
11. $D2 = N = C1 + 2 * C2 + 3 * C3 + 4 * C4$
12. $D3 = X = A2 * (C1) + 2 * A2 * (C2 + C3 + C4)$
13. $D4 = R = D2/D3$
14. $D5 = LC = A1 * C4$

Now open the SOLVER window. Do the following: (1) target cell = $D1$, or $D2, D3, D4, D5$ (what to max or min); (2) changing cells = $A1 : A2$ (the variables); and (3) constraints are $3 \leq A1 \leq 5$ and $5 \leq A2 \leq 7$ for the $\alpha = 0$ cut. Choose "max" or "min" and click the solve button. In the options box we used automatic scaling, estimates = tangent, derivatives = forward, and search = Newton. Table 7.2 has the optimization results.

The major problems with SOLVER are: (1) it can get out of the feasible set; and (2) it can stop at a local maximum. If Solver gives an error message (left the feasible set) just abort that run and start again. To guard against finding local maximums/minimums you will need to run SOLVER many times (the more the better) with different initial conditions.

Table 7.2. Optimal Values of $\overline{U}$, $\overline{N}$, $\overline{X}$, $\overline{R}$, $\overline{LC}$ ($c = 2$, $M = 4$)

	Min	λ	μ	Max	λ	μ
U	0.3525	3	7	0.6522	5	5
N	0.4456	3	7	1.1304	5	5
X	2.9737	3	5	4.9223	5	7
R	0.1489	3	7	0.2364	5	5
LC	0.0082	3	7	0.2174	5	5

7.2.2 Simulation

We have a simulation package (GPSS, Chap. 6) to simulate crisp queuing systems which we now call **SIM**. What **SIM** will do is

$$N = \mathbf{SIM}(inputs) , \tag{7.18}$$

where N is the estimated mean value of the number of customers in the system from the simulation, since the simulation produces a distribution of values for N, and $inputs = v = (c, M, \lambda, \mu, \ldots)$. In the inputs c is the number of parallel servers and M is system capacity. In our example $c = 2$ and $M = 4$. We can use simulation to estimate alpha-cuts of $\overline{N}$. Let $\overline{N}[\alpha] = [n_1(\alpha), n_2(\alpha)]$.

$$[n_1(\alpha), n_2(\alpha)] = \{\mathbf{SIM}(v) | \lambda \in \overline{\lambda}[\alpha], \mu \in \overline{\mu}[\alpha]\} , \tag{7.19}$$

for $0 \leq \alpha \leq 1$. So

$$n_1(\alpha) = min\{\mathbf{SIM}(v) | \lambda \in \overline{\lambda}[\alpha], \mu \in \overline{\mu}[\alpha]\} , \tag{7.20}$$

see (7.12), and

$$n_2(\alpha) = max\{\mathbf{SIM}(v) | \lambda \in \overline{\lambda}[\alpha], \mu \in \overline{\mu}[\alpha]\} , \tag{7.21}$$

see (7.13). Hence, from SOLVER in Example 7.2.1.1 we know the values of λ_l and μ_l in (7.16) so

$$n_1(\alpha) = \mathbf{SIM}(2, 4, \lambda_l = 3, \mu_l = 7) , \tag{7.22}$$

and we also know the values of λ_u and μ_u in (7.17) so

$$n_2(\alpha) = \mathbf{SIM}(2, 4, \lambda_u = 5, \mu_u = 5) . \tag{7.23}$$

This means that simulation can estimate the end points of the alpha-cuts of $\overline{N}$. Similar remarks hold for $\overline{U}$, $\overline{X}$, $\overline{R}$ and $\overline{LC}$.

Example 7.2.2.1

This continues Example 7.2.1.1. We used GPSS/H. The simulation program
was

```
1.  SIMULATE
2.  REAL  &C(1),&D(1),&E(1),&F(1)
3.  OUT  FILEDEF 'ANS.OUT'
4.  STORAGE  S(WORK),2
5.  STORAGE  S(BUFFER),2
6.  GENERATE  RVEXPO(1,0.3333)
7.  TEST LE  S(BUFFER),1,POUT2
8.  QUEUE SYSTQ
9.  TEST E  S(WORK),0,JUMP1
10. BLET  &D(1)=&D(1)+1
11. TRANSFER ,NEXT
12. JUMP1  TEST E  S(WORK),1,JUMP2
13. BLET  &F(1)=&F(1)+1
14. TRANSFER ,NEXT
15. JUMP2  TEST E  S(WORK),2,NEXT
16. BLET  &C(1)=&C(1)+1
17. NEXT  ENTER  BUFFER
18. ADVANCE
19. ENTER  WORK
20. LEAVE  BUFFER
21. ADVANCE  RVEXPO(1,0.1428)
22. LEAVE  WORK
23. DEPART SYSTQ
24. TRANSFER ,POUT1
25. POUT2  BLET  &E(1)=&E(1)+1
26. TERMINATE 1
27. POUT1  TERMINATE  1
28. START  500,000
29. PUTPIC  FILE=OUT,LINES=8,(&C(1),&F(1),&D(1),&E(1))
    NUMBER FULL ****
    NUMBER ONE ****
    NUMBER EMPTY ****
    NUMBER LOST ****
30. END
```

A brief explanation of this program is needed. The command "SIMU-
LATE" starts it off and then we define &C(1),&D(1),&E(1) and &F(1) to
be a real numbers. &C(1) will count the number of times the servers are
both busy, &D(1) counts the number of times both servers are idle, &E(1)
counts the number of lost customers due to the system at capacity and &F(1)
counts the number of times only one server is busy. The next line defines a

file "OUT" which will contain the final value of $\&C(1),\&D(1),\&E(1)$ and $\&F(1)$. Two storage items are defined next: (1) WORK will contain the two identical and parallel servers; and (2) BUFFER will be the queue in front of WORK with capacity two customers. GENERATE makes arrivals according to the exponential distribution, using random number stream number one, with mean 0.3333 time units between arrivals. The TEST statement sees if the system is at capacity (4), and if it is the customer is sent to POUT2, counted as a lost customer and TERMINATED. QUEUE SYSQ, and later DEPART SYSQ, allows for statistics (time in the system,...) to be gathered on the system. Then we have a series of tests to update the values of $\&C(1),\&D(1)$ and $\&F(1)$. The customer then enters BUFFER if the servers are busy and ADVANCE, all by itself, says there is no delay to be experienced within BUFFER except to wait for a free server. When a server is free a customer can enter WORK and the server time is governed by the exponential with mean 0.1428. Finally, after WORK, the customer goes to POUT1 and leaves the system. The file OUT will contain the number of times the system was empty when a new customer arrived, the total number of lost customers, etc. The rest of the statistics we want will be contained in the standard output report compiled by GPSS/H. The author does not claim that these GPSS programs are the most efficient since he learned the language while writing this book.

We ran the program until 500,000 customers arrived at the system. Run time would be expected to be less than 5 seconds. There are two common methods of handling "steady-state" statistics [1]: (1) "swamping"; and (2) "deletion". In the exact one-step calculations we assumed that we were in steady-state. Initially at start-up, the system has some transient behavior, due to it beginning empty, and as time goes on the transient behavior dies down and we proceed into steady-state. In the "swamping" method we simulate for a long period so that the effect of the initial conditions becomes insignificant. This is what we did with a simulation of 500,000 customers. In the "deletion" method we would break the simulation down into two parts, an initial warm up phase, followed by a data collection period. We will use the swamping method throughout the book.

The results of the simulation, using the optimal values of λ and μ from Table 7.2, are shown in Table 7.3. All of these values are very close to those in Table 7.2.

In the standard output report from GPSS/H we find N and R under QUEUE. To estimate p_0 we compute $\&D(1)/500,000$ and then $1 - p_0$ is U. Some customers are lost due to maximum capacity of 4, but the standard output report contains CLOCK = the total number of elapsed time units for the 500,000 simulation run. Then $500,000 - \&E(1)$ = the total number of customers who left the system after passing through a server. Hence, X is estimated by $[500,000 - \&E(1)]/CLOCK$. LC is estimated by $\&E(1)/CLOCK$.

Table 7.3. Simulation Values of $\overline{U}, \overline{N}, \overline{X}, \overline{R}, \overline{LC}$ ($c = 2, M = 4$)

	Min	λ_l	μ_l	Max	λ_u	μ_u
U	0.3521	3	7	0.6521	5	5
N	0.444	3	7	1.133	5	5
X	2.9702	3	5	4.9168	5	7
R	0.149	3	7	0.237	5	5
LC	0.0083	3	7	0.2181	5	5

In general, we may not know the "one-step" functions, or we may not know the optimal values of λ and μ, but simulation can still be used to estimate the intervals for the α-cuts of $\overline{N}$ by solving (7.20) and (7.21). We might need to employ some optimization procedure (see Chap. 8) for the min (max) in (7.20) (7.21). What we do to use simulation to estimate the alpha-cuts of $\overline{N}$ will also be employed on $\overline{U}, \overline{X}, \overline{R}$ and $\overline{LC}$.

7.3 Multi-Step Calculations

In this section we argue that if we compute $\overline{N}$ (or $\overline{U}, \overline{X}, \overline{R}, \overline{LC}$) from a series of fuzzy calculations (at least two), then the $\overline{N}$ we get may be more fuzzy than what one obtains from the one-step procedure. It is also possible that the two methods produce the same result. Also, crisp simulation may not approximate alpha-cuts of the multi-step $\overline{N}$.

We continue with the same example used in the previous sections. Given λ and μ we could first find the steady-state probabilities p_i, $0 \leq i \leq 4$. Formulas for the p_i were given in Sect. 7.2. In this section assume that

$$p_i = f_i(\lambda, \mu), \tag{7.24}$$

$0 \leq i \leq 4$. Then there will be formulas for N (and U, X, R, LC) in terms of the p_i. For example

$$N = p_1 + 2p_2 + 3p_3 + 4p_4. \tag{7.25}$$

Now consider fuzzy arrival rate $\overline{\lambda}$ and fuzzy service rate $\overline{\mu}$. We have a two-step procedure. First we use the extension principle to get fuzzy steady-state probabilities

$$\overline{p}_i = f_i(\overline{\lambda}, \overline{\mu}), \tag{7.26}$$

$0 \leq i \leq 4$. To compute $\overline{N}[\alpha]$ we use restricted fuzzy arithmetic [2–5].

$$\overline{N}[\alpha] = \left\{ \sum_{i=0}^{4} (i)p_i | \mathbf{S} \right\}, \tag{7.27}$$

where $\mathbf{S}$ is the statement "$p_i \in \overline{p}_i[\alpha], 0 \leq i \leq 4, p_0 + p_1 + \cdots + p_4 = 1$". $\overline{p}_i$, $0 \leq i \leq 4$, is a discrete fuzzy probability distribution. We have uncertainty about the values of the p_i but there is no uncertainty that we have a discrete probability distribution. Hence, we may choose the p_i in the alpha-cuts of the fuzzy probabilities but the sum must equal one. There are values p_{il} and p_{iu} in $\overline{p}_i[\alpha]$, $0 \leq i \leq 4$, so that

$$n_1(\alpha) = \sum_{i=0}^{4}(i)p_{il} \, , \tag{7.28}$$

$$n_2(\alpha) = \sum_{i=0}^{4}(i)p_{iu} \, . \tag{7.29}$$

The p_{il}, $0 \leq i \leq 4$, give the left end point of $\overline{N}[\alpha]$ and the p_{iu}, $0 \leq i \leq 4$, produce the right end point. Now there are also $\lambda_{il} \in \overline{\lambda}[\alpha]$ and $\mu_{il} \in \overline{\mu}[\alpha]$ so that

$$p_{il} = f_i(\lambda_{il}, \mu_{il}) \, , \tag{7.30}$$

$0 \leq i \leq 4$. Also there are $\lambda_{iu} \in \overline{\lambda}[\alpha]$ and $\mu_{iu} \in \overline{\mu}[\alpha]$ so that

$$p_{iu} = f_i(\lambda_{iu}, \mu_{iu}) \, , \tag{7.31}$$

$0 \leq i \leq 4$. The end result is that for this method of computation we use a maximum of 5 values of $\lambda \in \overline{\lambda}[\alpha]$ and a maximum of 5 values of $\mu \in \overline{\mu}[\alpha]$ to find the left end point of $\overline{N}[\alpha]$. Similar results for the right end point of $\overline{N}[\alpha]$. If we let $\overline{N}^{(2)}$ be the fuzzy number we obtain from the two-step calculation and $\overline{N}^{(1)}$ from the one-step method, we see that $\overline{N}^{(1)} \leq \overline{N}^{(2)}$ (see Sect. 2.2.3). That is, $\overline{N}^{(1)}$ may be less fuzzy than $\overline{N}^{(2)}$, but they could be equal. Since simulation approximates the one-step method, simulation may not approximate the two-step (or three-step, etc.) calculation.

Example 7.3.1

This continues the previous examples in this chapter with $\overline{\lambda} = (3/4/5)$ and $\overline{\mu} = (5/6/7)$. We compute the $\alpha = 0$ cuts of $\overline{U}, \overline{N}, \overline{X}, \overline{R}$ and $\overline{LC}$ using the two-step method.

First, using SOLVER, we find the $\alpha = 0$ cuts of the fuzzy steady-state probabilities. The results are in Table 7.4.

Next we input the $\overline{p}_i$, $0 \leq i \leq 4$, into the expressions for the U, N, X, R, and LC, (7.2)–(7.6), and compute $\overline{U}, \ldots, \overline{LC}$ using the extension principle with final results given in Table 7.5. Notice in Table 7.5 that the right end point of $\overline{X}[0]$ equals 5.0000 because $\overline{X} \leq 5 =$ the maximum arrival rate. The actual value computed here was 6.6962. So, the fuzzy number $\overline{X}$ gets cut off at 5.0000.

Now compare Tables 7.1 (one-step results) to Table 7.4 (two-step results). We see that $\overline{X}^{(1)} \leq \overline{X}^{(2)}$ and $\overline{Z}^{(1)} \approx \overline{Z}^{(2)}$ for $\overline{Z} \in \{\overline{U}, \overline{N}, \overline{R}, \overline{LC}\}$.

Table 7.4. Alpha Cuts of the Fuzzy Probabilities in Example 7.3.1 ($c = 2$, $M = 4$)

	Alpha = Zero Cut	Alpha = One Cut
$\overline{p}_0$	$[0.3478, 0.6475]$	0.5031
$\overline{p}_1$	$[0.2775, 0.3503]$	0.3354
$\overline{p}_2$	$[0.0595, 0.1739]$	0.1118
$\overline{p}_3$	$[0.0127, 0.0870]$	0.0373
$\overline{p}_4$	$[0.0027, 0.0435]$	0.0124

Table 7.5. Two-Step Alpha Cuts of $\overline{U}$, $\overline{N}$, $\overline{X}$, $\overline{R}$, $\overline{LC}$ in Example 7.3.1 ($c = 2$, $M = 4$)

	Alpha = Zero Cut	Alpha = One Cut
$\overline{U}$	$[0.3525, 0.6522]$	0.4969
$\overline{N}$	$[0.4455, 1.1306]$	0.7205
$\overline{X}$	$[2.1370, 5.0000]$	3.9504
$\overline{R}$	$[0.1464, 0.2529]$	0.1824
$\overline{LC}$	$[0.0081, 0.2175]$	0.0497

Example 7.3.1 shows that in using two-step fuzzy calculations sometimes we get the same result as in the one-step calculation and we may get a different (more fuzzy) answer. In the rest of the book we will want to compute many other fuzzy numbers besides those for U, N, X, R and LC. We will usually not know, for all these fuzzy numbers, if the one-step and the multi-step fuzzy calculations agree. So, to be on the "safe side" we will only use simulation to approximate the α-cuts of the fuzzy numbers obtained from one-step calculations.

7.4 Rest of the Book

In the applications in the following chapters, Chaps. 9–26, we will use simulation to approximate alpha-cuts of certain fuzzy system descriptors like $\overline{N}$, $\overline{U}$, $\overline{X}$, $\overline{R}$ and $\overline{LC}$ when we do not have a one-step calculation available. These systems will be sufficiently complicated so that a one-step expression for N, U, X, R and LC is too difficult to derive and we do not know how to find it in the literature. We assume a continuous one-step function theoretically exists, but in practice we do not know a formula for this function, and the simulation approximates the results of this (unknown) one-step calculation. The way to analyze these fuzzy systems is through simulation.

All of these systems use probabilities and probability distributions. We must estimate these probabilities, or the parameters in the distributions, from crisp data. Our estimators will all be fuzzy estimators (Chap. 3). Fuzzy estimators force the system to become a fuzzy system whose analysis is accomplished by computing fuzzy numbers for $\overline{N}$, $\overline{U}$, $\overline{X}$, $\overline{R}$, etc. All these fuzzy calculations will be approximated by simulation.

References

1. J. Banks, J.S. Carson and J.N. Sy: Getting Started with GPSS/H, Second Edition, Wolverine Software Corp., Annandale, VA, 1995.
2. J.J. Buckley: Fuzzy Probabilities: New Approach and Applications, Physica-Verlag, Heidelberg, Germany, 2003.
3. J.J. Buckley: Fuzzy Probabilities and Fuzzy Sets for Web Planning, Springer, Heidelberg, Germany, 2004.
4. J.J. Buckley and E. Eslami: Uncertain Probabilities I: The Discrete Case, Soft Computing, 7(2003)500-505.
5. J.J. Buckley, K. Reilly and X. Zheng: Fuzzy Probabilities for Web Planning, Soft Computing, 8(2004)464-476.
6. J.J. Buckley: Maximum Entropy Principle with Imprecise Side-Conditions II: Crisp Discrete Solutions, Soft Computing. To appear.
7. Frontline Systems (www.frontsys.com, www.solver.com).
8. H.A. Taha: Operations Research, Fifth Edition, Macmillan, N.Y., 1992.

8 Simulation Optimization

8.1 Introduction

In this chapter we discuss how we plan to solve the optimization problems attached to the simulation as expressed in (5.4)–(5.5) in Chap. 5, or (7.20)–(7.21) in Chap. 7, without the use of a one-step function. Consider the fuzzy system of Chap. 5, shown in Fig. 5.1, and the servers S41,S42,S43. Each of these servers has service time described by a normal distribution $N(\mu_4, \sigma_4^2)$. For this fuzzy system we want to compute $\overline{X}$ the fuzzy number for throughput. To approximate $\overline{X}$ we need some of its α-cuts. Now we first obtain fuzzy estimators for μ_4 and σ_4^2 which are $\overline{\mu}_4$ and $\overline{\sigma}_4^2$, respectively. Choose $\alpha \in [0, 1)$. The optimization problem to solve is to decide on $\mu \in \overline{\mu}_4[\alpha]$ and $\sigma \in \overline{\sigma}_4^2[\alpha]$ to use in the simulation so that the output approximates the end points of $\overline{X}[\alpha]$. This has to be done for all the fuzzy estimators in all the fuzzy distributions employed by the system.

In this book there are only three places in a fuzzy system that depend on a fuzzy distribution: (1) arrivals; (2) service stations; and (3) random transfer points. We now look at each of these and how we hope to solve the optimization problem.

8.2 Arrivals

In this book all arrivals will usually be governed by the exponential distribution. To describe arrivals we use $\lambda = $ the number of items (customers) arriving per unit time and $1/\lambda = $ the mean time between arrivals. The solution to the optimization problem is given in Table 8.1. We have assumed that all parameters in the fuzzy system are held fixed except λ for this arrival. If $\overline{\lambda}$ is our fuzzy estimator of λ, let $[\lambda_l, \lambda_u] = [\lambda_1(\alpha), \lambda_2(\alpha)] = \overline{\lambda}[\alpha]$ for some fixed $\alpha \in [0, 1)$.

We interpret Table 8.1 as follows: (1) if we want to approximate the left end point of the interval $\overline{X}[\alpha]$ use for λ the left end point of the interval $\overline{\lambda}[\alpha]$; and (2) if you want to approximate the right end point of the interval $\overline{R}$ use for λ the right end point of the interval $\overline{\lambda}[\alpha]$.

All of these results are rather intuitive. Let λ increase from its minimum λ_l to its maximum λ_u. All other parameters of the system are held fixed. The

James J. Buckley: *Simulating Fuzzy Systems*, StudFuzz **171**, 69–73 (2005)
www.springerlink.com   © Springer-Verlag Berlin Heidelberg 2005

Table 8.1. Optimal Values of $\lambda \in [\lambda_l, \lambda_u] = \overline{\lambda}[\alpha]$ for Fuzzy Exponential Arrivals in Simulation

Min	λ	Max	λ
U	λ_l	U	λ_u
N	λ_l	N	λ_u
R	λ_l	R	λ_u
X	λ_l	X	λ_u
LC	λ_l	LC	λ_u

system becomes more congested. We will call arrivals "customers", but they need not be people. More and more customers will fill the queues in front of the servers. This means that server utilization (mean value $= U$) increases, number of customers in the system (mean value $= N$) increases and the time a customer spends in the system (mean value $= R$) also increases. At first it is not clear what happens to the number of customers leaving the system per unit time (mean $= X$) since $X = N/R$. But the number leaving per unit time will not decrease, so X will also increase. If the system has finite capacity, then the number of customers lost per unit time (LC) increases as λ increases. These results are what is in Table 8.1. The same results are shown in Tables 7.2 and 7.3. The units in Table 8.1 for λ are number of customers arriving per unit time and not the number of time units between arrivals.

8.3 Service

In this book we will use for the service probability distribution the: (1) exponential; (2) uniform; and (3) normal.

8.3.1 Exponential

The number of service completions per unit time is μ so the mean service time is $1/\mu$. The exponential depends only on one parameter μ. The solution to the optimization problem is given in Table 8.2. If $\overline{\mu}$ is our fuzzy estimator of μ, let $[\mu_l, \mu_u] = [\mu_1(\alpha), \mu_2(\alpha)] = \overline{\mu}[\alpha]$ for some fixed $\alpha \in [0, 1)$.

We interpret Table 8.2 as follows: (1) if we want to approximate the left end point of the interval $\overline{X}[\alpha]$ use for μ the left end point of the interval $\overline{\mu}[\alpha]$; and (2) if you want to approximate the right end point of the interval $\overline{R}$ use for μ the left end point of the interval $\overline{\mu}[\alpha]$.

All of these results are also intuitive. Let μ increase from its minimum μ_l to its maximum μ_u. All other parameters of the system are held fixed. This service station may have c ($c \geq 1$) parallel and identical servers. The system consisting of this server and what comes before it, becomes less congested. Fewer and fewer customers fill the queue in front of this server. This means

Table 8.2. Optimal Values of $\mu \in [\mu_l, \mu_u] = \overline{\mu}[\alpha]$ for Fuzzy Exponential Service in Simulation

Min	μ	Max	μ
U	μ_u	U	μ_l
N	μ_u	N	μ_l
R	μ_u	R	μ_l
X	μ_l	X	μ_u
LC	μ_u	LC	μ_l

that server utilization (mean value $= U$) decreases, the number of customers in the system (mean value $= N$) decreases and the time a customer spends in the system (mean value $= R$) also decreases. It is not clear what happens to the number of customers leaving the system per unit time (mean $= X$) since $X = N/R$. But the number leaving per unit time will not decrease, so X will increase. If the system has finite capacity, then the number of customers lost per unit time (LC) decreases as μ increases. These results are what is in Table 8.2. The same results are shown in Tables 7.2 and 7.3. The units in Table 8.2 are the number of service completions per unit time and not the number of time units between service completions.

8.3.2 Uniform

The uniform distribution $U(a,b)$ depends on two parameters a and b. Our fuzzy estimator for a is $\overline{a}$ and for b we use $\overline{b}$. See Figs. 3.8 and 3.9. The solution to the optimization problem is in Table 8.3. Pick and fix $\alpha \in [0,1)$. Let $\overline{a}[\alpha] = [a_l, a_u]$ where $a_l \geq 0$, and a_u is fixed for all α and let $\overline{b}[\alpha] = [b_l, b_u]$ where b_l is fixed for all α (see Figs. 3.8 and 3.9).

We interpret Table 8.3 as follows: (1) if we want to approximate the left end point of the interval $\overline{X}[\alpha]$ use for a the right end point of the interval $\overline{a}[\alpha]$ and for b use the right end point of the interval $\overline{b}[\alpha]$; and (2) if you want to approximate the left end point of the interval $\overline{R}$ use for a the left end point of the interval $\overline{a}[\alpha]$ and for b use the left end point of the interval $\overline{b}[\alpha]$.

Table 8.3. Optimal Values of a and b for Fuzzy Uniform Service in Simulation

Min	a	b	Max	a	b
U	a_l	b_l	U	a_u	b_u
N	a_l	b_l	N	a_u	b_u
R	a_l	b_l	R	a_u	b_u
X	a_u	b_u	X	a_l	b_l
LC	a_l	b_l	LC	a_u	b_u

It can be confusing to compare Tables 8.2 and 8.3. Table 8.2 uses max/min rates but Table 8.3 has max/min times. A min (max) rate corresponds to a max (min) time. The values in Table 8.3 are about times to service completion.

The discussion on why Table 8.3 is correct is similar to why Table 8.2 is correct. All other parameters are held fixed. Analysis of Table 8.3 centers around the mean service time $((a + b)/2)$ of the uniform $U(a, b)$. You minimize $(a + b)/2$ to minimize U, N, R, LC and you maximize $(a + b)/2$ to maximize U, N, R, LC. But you minimize (maximize) the mean service time to maximize (minimize) X.

8.3.3 Normal

The normal distribution $N(\mu, \sigma^2)$ depends on two parameters μ and σ^2. Our fuzzy estimator for μ is $\overline{\mu}$ and for σ^2 we use $\overline{\sigma}^2$. See Chap. 3. The solution to the optimization problem is in Table 8.4. Pick and fix $\alpha \in [0, 1)$. Let $\overline{\mu}[\alpha] = [\mu_l, \mu_u]$.

Table 8.4. Optimal Values of μ for Fuzzy Normal Service in Simulation

Min	μ	Max	μ
U	μ_l	U	μ_u
N	μ_l	N	μ_u
R	μ_l	R	μ_u
X	μ_u	X	μ_l
LC	μ_l	LC	μ_u

We interpret Table 8.4 as follows: (1) if we want to approximate the left end point of the interval $\overline{X}[\alpha]$ use for μ the right end point of the interval $\overline{\mu}[\alpha]$; and (2) if you want to approximate the left end point of the interval $\overline{R}$ use for μ the left end point of the interval $\overline{\mu}[\alpha]$. The values in Table 8.4 are service times not service rates.

The discussion on why Table 8.3 is correct is similar to why Tables 8.2 and 8.3 are correct. We pick and fix the values of all the other parameters. You minimize (maximize) μ to minimize (maximize) U, N, R, LC and you maximize (minimize) μ to minimize (maximize) X.

Notice that no values of σ^2 are in Table 8.4. This is because the results are approximately independent of σ^2. If you hold μ fixed and vary σ^2 throughout some interval, remember to keep $\mu - 5\sigma^2 > 0$, the values for U, N, X, R, LC are relatively constant. We did notice that the number of customers lost can increase as σ^2 increases, but the value of the CLOCK can also change, leaving LC was almost constant. All simulations using normal service times, showing independence of the variance, were done using the example in Chap. 7 (Fig. 7.1).

So, in the rest of this book if σ^2 is unknown we will use its point estimator (where the membership value is one in the fuzzy estimator) in all simulations.

8.4 Probabilistic Transfer

A probabilistic transfer in a system is like that in the example in Chap. 5, see Fig. 5.1, where the item goes to "Reject" with probability p and to "Accept" with probability $1-p$. Assume that the value of p is not known precisely and must be estimated from data. Then we obtain a fuzzy estimator $\overline{p}$ of p and $\overline{q} = 1 - \overline{p}$ for $1 - p$.

Pick and fix $\alpha \in [0, 1)$. Let $[p_l, p_u] = \overline{p}[\alpha]$ and $[q_l, q_u] = \overline{q}[\alpha]$. If $p \in [p_l, p_u]$, then $q = 1 - p$ is in $[q_l, q_u]$. Now we want to know how to choose p in $[p_l, p_u]$ to maximize/minimize U, N, X, R and LC. We hold the values of all the other parameters fixed and it will depend on where the transfer point is within the network.

Sometimes it is obvious on how to choose p. In Fig. 5.1 assume we wish to maximize X = mean number of accepted items leaving the system per unit time. Then pick p to maximize $1 - p$, or $p = p_l$. If it is not absolutely clear on how to choose p, then we will experiment with the simulation by choosing a spread of p values in $[p_l, p_u]$ to find out the max/min of U, N, etc.

8.5 Summary

Consider the network in Fig. 5.1 where many of the parameters are estimated from data and we therefore get fuzzy estimators. Approximate the fuzzy numbers $\overline{U}$, $\overline{N}$, $\overline{R}$, $\overline{X}$ and $\overline{LC}$. We will do this approximation only through the $\alpha = 0, 0.5, 1$ cuts. Tables 8.1–8.4 tell us how to set the values for all the parameters, except p, to obtain these approximations and it is obvious (for Fig. 5.1) on what to do with p. The next chapter does the simulation for the fuzzy system in Fig. 5.1.

Whenever we use the fuzzy normal for service times the estimator of the mean will be a fuzzy number $\overline{\mu}$ but the estimator for the variance will always be crisp.

As we go through the rest of the book there will be many fuzzy system descriptors to be estimated using simulation besides U, N, X, R and LC. We will construct tables like Tables 8.1–8.4 for each new fuzzy system descriptor when they are needed.

9 Queuing II: No One-Step Calculations

9.1 Introduction

In this chapter we will study the fuzzy system shown in Fig. 5.1 now re-produced as Fig. 9.1. This example was adapted from an example in [1]. The fuzzy probability distributions we will use in the system are defined in Table 9.1. There have been some changes in the assignment of probability distributions from that done in Chap. 5. Arrivals occur according to the fuzzy exponential distribution with mean $1/\overline{\lambda}$ for fuzzy estimator $\overline{\lambda} = (3/4/5)$. We first have that the operations at S1 and S3 are "robotic" (robots) and we know the service time to be uniform given in Table 9.1. Next, the operations performed at S2, S4 and S5 are done by workers plus machines and service time can vary and needs to be estimated. Their service time is modelled by the normal distribution with fuzzy estimator for the mean. However, from the discussion in the previous chapter, we only use a point estimator for the standard deviation (variance). Finally, the probabilistic transfer point at the end of the network (accept/reject) also has to be estimated from data and we obtain the fuzzy number estimator $\overline{p}$ for p. For simplicity, all our fuzzy numbers will be triangular fuzzy numbers.

Let us assume, for another change from Chap. 5, that all queues have unlimited storage with queue discipline "first come , first served" (no priorities). The goal is to find the fuzzy numbers $\overline{N}$, $\overline{X}$ and $\overline{R}$ and consider some simple changes in the network so as to increase $\overline{X}$ = throughput. Assume the time units are in minutes. In this chapter X will be the number of accepted units per 8 hour shift (day). For a minimal approximation to these fuzzy numbers we will use simulation to approximate the $\alpha = 0, 0.5, 1$ cuts. One-step formulas for N, X and R would be complicated and difficult to derive, so we omit them and go directly to crisp simulation (see Chap. 7).

9.2 Case 1: First Simulation

Using Chap. 8 we know how to choose λ, μ_2, μ_4, μ_{51} and μ_{52} to approximate the end points of α-cuts of $\overline{N}$, $\overline{X}$ and $\overline{R}$. Tables 9.2 and 9.3 has these results for the $\alpha = 0$ cut. Now as p varies in the interval $[0.03, 0.07]$ the values of N

James J. Buckley: *Simulating Fuzzy Systems*, StudFuzz **171**, 75–80 (2005)
www.springerlink.com

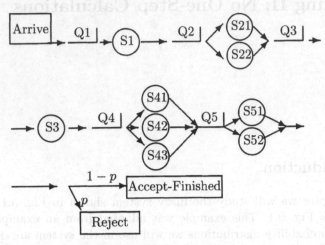

Fig. 9.1. Queuing System

and R do not change, but max (min) X will be for min (max) p. The GPSS program for this case is in Chap. 28.

Obvious changes to estimate the $\alpha = 0.5$ cut. Finally, we just use the $\alpha = 1$ cut values of the parameters for the alpha equal one cut of $\overline{N}$, $\overline{X}$ and $\overline{R}$. Results are in Table 9.4. In this chapter all simulations were run until 10,000 items were "accepted" (not rejected). The simulation output gives the value of CLOCK which is the total minutes for the simulation. A typical value

Table 9.1. Fuzzy/Crisp Probability Distributions for Fig. 9.1

Item	Distribution	Details
Arrivals	Exponential	$\overline{\lambda} = (3/4/5)$
Server = S1	Uniform	Crisp = $[0.14, 0.20]$
Servers = S2	Normal	$\overline{\mu}_2 = (0.1/0.15/0.2), \sigma_2 = 0.02$
Server = S3	Uniform	Crisp = $[0.14, 0.20]$
Servers = S4	Normal	$\overline{\mu}_4 = (0.1/0.15/0.2), \sigma_4 = 0.02$
Server = S51	Normal	$\overline{\mu}_{51} = (0.05/0.1/0.15), \sigma_{51} = 0.01$
Server = S52	Normal	$\overline{\mu}_{52} = (0.15/0.20/0.25), \sigma_{52} = 0.03$
Transfer	Bernoulli	$\overline{p} = (0.03/0.05/0.07)$

Table 9.2. Alpha Zero Cut Values of the Parameters for Min N, X, R

Min	λ	μ_2	μ_4	μ_{51}	μ_{52}	p
N	3	0.1	0.1	0.05	0.15	any p
X	3	0.2	0.2	0.15	0.25	0.07
R	3	0.1	0.1	0.05	0.15	any p

Table 9.3. Alpha Zero Cut Values of the Parameters for Max N, X, R

Max	λ	μ_2	μ_4	μ_{51}	μ_{52}	p
N	5	0.2	0.2	0.15	0.25	any p
X	5	0.1	0.1	0.05	0.15	0.03
R	5	0.2	0.2	0.15	0.25	any p

Table 9.4. Case 1 Simulation Alpha-Cuts of $\overline{N}$, $\overline{X}$, $\overline{R}$

Item	$\alpha = 0$ Cut	$\alpha = 0.5$ Cut	$\alpha = 1$ Cut
N	[2.081, 7.071]	[2.829, 5.056]	3.680
X	[1349, 2299]	[1576, 2038]	1791
R	[0.698, 1.420]	[0.813, 1.142]	0.938

for CLOCK is 1902 which is almost four 8 hour days of simulation and the execution time was less than one second. So $X = (10,000/CLOCK)(60)(8)$. In the simulation output R is the mean time a unit spends in the system from arrival till the final inspection is completed (includes both accepted and rejected items). Also, from the standard simulation printout N is the average number of units in the system over the whole simulation.

Throughout this book we employ the "swamping" method, discussed in Chap. 7, to minimize the effect of staring the simulation when all queues/servers are empty. The swamping method is to run the simulation long enough to be sure that it has been in steady-state long enough so that the start up effects are minimized. Running for 10,000 accepted units should be sufficient.

We show the graphs of $\overline{N}$ and $\overline{R}$ in Figs. 9.2 and 9.3, respectively. $\overline{X}$ is in Fig. 9.4 together with $\overline{X}$ from the next two simulations.

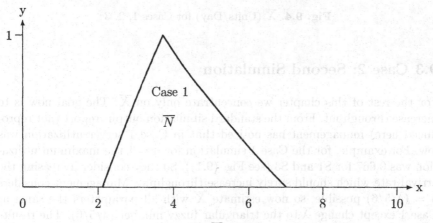

Fig. 9.2. $\overline{N}$ in Case 1

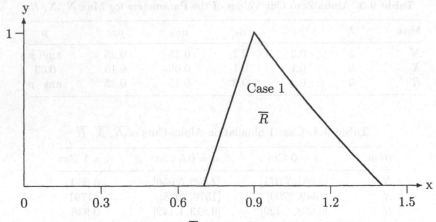

Fig. 9.3. $\overline{R}$ (Minutes) in Case 1

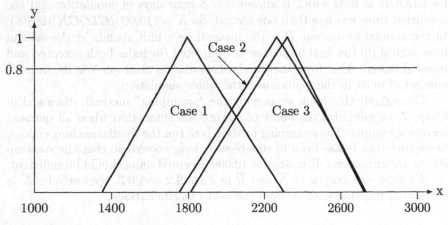

Fig. 9.4. $\overline{X}$ (Units/Day) for Cases 1, 2, 3

9.3 Case 2: Second Simulation

For the rest of this chapter we concentrate only on $\overline{X}$. The goal now is to increase throughput. From the standard simulation output report (not reproduced here) management has noticed that in Case 1 server utilization was low. For example, for the Case 1 simulation for $\alpha = 1$ the maximum utilization was 0.667 for S1 and S3 (see Fig. (9.1)). So they consider increasing the arrival rate which should surely increase throughput. Management considers $\overline{\lambda} = (4/5/6)$ possible, so now estimate $\overline{X}$ with all parameters the same as Case 1 except change $\overline{\lambda}$ to the triangular fuzzy number $(4/5/6)$. The results are in Table 9.5 and the graph of Case 2 $\overline{X}$ is in Fig. 9.4.

Table 9.5. Case 2 Simulation Alpha-Cuts of $\overline{X}$

Item	$\alpha = 0$ Cut	$\alpha = 0.5$ Cut	$\alpha = 1$ Cut
X	[1748, 2728]	[2007, 2532]	2264

We see that throughput has definitely increased and so has server utilization. In the simulation for $\alpha = 1$ maximum utilization has increased to 0.884 for S1 and S3.

9.4 Case 3: Third Simulation

Management has been thinking about replacing the workers at S2 and S4 (see Fig. 9.1) with robots. S2 will have two robotic work stations and S4 will have three robotic work stations. Service time then at S2 and S4 will be a crisp uniform distribution for each robotic server. Also, with all robotic servers the percent of rejects at the end of the production line should decrease. A new value of $\overline{p}$ is estimated to be (0.03/0.04/0.05). From the standard simulation output report management has noticed that the second inspector (slower, less experienced) S52 is not being sufficiently utilized and they propose eliminating the second inspector. Now simulate the system and estimate $\overline{X}$. Everything is the same as Case 2 except: (1) the probability distribution for S2 and S4 is $U(0.14, 0.20)$; (2) there is only one inspector S51; and (3) $\overline{p} = (0.03/0.04/0.05)$. The results for $\overline{X}$ are displayed in Table 9.6 with a graph for Case 3 $\overline{X}$ in Fig. 9.4. Also, checking the simulation output report we see that the single inspector's utilization is not excessive so the system does not miss the removal of the other inspector.

Table 9.6. Case 3 Simulation Alpha-Cuts of $\overline{X}$

Item	$\alpha = 0$ Cut	$\alpha = 0.5$ Cut	$\alpha = 1$ Cut
X	[1809, 2726]	[2051, 2522]	2318

Now we go back and review Sect. 2.5 in Chap. 2 on our method of ranking fuzzy numbers. Let $\overline{X}_i$ = throughput for Case i, $1 \leq i \leq 3$. We see from Fig. 9.4 that $\overline{X}_1 < \overline{X}_2 \approx \overline{X}_3$ and Cases 2 and 3 produce approximately the same fuzzy throughput. $\overline{X}$ is a discrete fuzzy set whose graph is approximated by a continuous fuzzy set (Sect. 2.7).

9.5 Summary

Management may go on with "what ifing" to increase $\overline{X}$. What if we add this? What if we change that? But we shall stop here. We used crisp simulation to approximate the one-step fuzzy numbers for $\overline{N}$, $\overline{X}$ and $\overline{R}$. These fuzzy numbers incorporate the uncertainty in the data and their bases are like a 99% confidence interval. We introduced the beginning of fuzzy system optimization. As you vary the fuzzy system you compute approximations to the fuzzy numbers describing performance and then you choose the configuration to possibly maximize throughput and/or minimize response time. See Sect. 2.6 of Chap. 2. We will continue to add some optimization concepts in the following chapters.

Reference

1. J. Banks, J.S. Carson and J.N. Sy: Getting Started with GPSS/H, Second Edition, Wolverine Software Corp. Annandale, VA, 1995.

10 Call Center Model

10.1 Introduction

The queuing system in this chapter is shown in Fig. 10.1. This application
was adopted from an example in [1].

When a call arrives at the call center the caller first listens to a recording,
chooses their call to be of type A, B or C, and then goes into the common
queue, assuming this waiting line is not at capacity. Let M denote queue
capacity. If the queue is at capacity M the caller gets a busy signal and
hangs up becoming a "lost customer". Let $\lambda_a(\lambda_b, \lambda_c)$ be the arrival rate for
calls of type A (B, C). Time units will be in minutes.

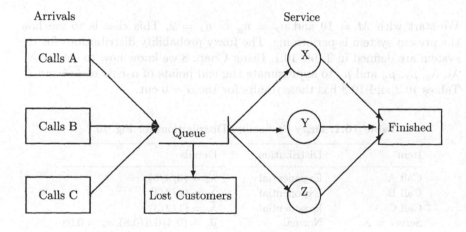

Fig. 10.1. Call Center Model

Once in the queue calls of type C can be handled only by server Z. Assume
each server X, Y, Z has n_x, n_y, n_z, respectively, parallel and identical servers
(operators). If all the operators in server Z are busy, then call C must wait
in the queue. A call of type B can be taken by server Y or Z. Call B first
tries to get an operator in server Y, and if all of these are busy, then goes
on to server Z, and when all the operators in Z are busy it must wait in the

James J. Buckley: *Simulating Fuzzy Systems*, StudFuzz **171**, 81–85 (2005)
www.springerlink.com   © Springer-Verlag Berlin Heidelberg 2005

queue. Lastly, a call of type A can be processed by any server but it starts at server X, goes next to Y when all of X is busy, and finally to Z if it can not be served by Y. Call A waits in the queue if X, Y and Z are busy. The service times for all the operators will be modelled by the normal distribution.

Management is interested in studying lost customers $\overline{LC}$ and average time a completed call spends in the system $\overline{R}$. The goal is to consider configurations to minimize both $\overline{LC}$ and $\overline{R}$. We have no one-step formulas for LC and R (see Chap. 7) so we will simulate the fuzzy system to estimate the $\alpha = 0, 0.5, 1$ cuts of $\overline{LC}$ and $\overline{R}$. All simulations will be until 10,000 calls are processed. The value of 10,000 seems sufficiently large to get the system well into steady-state so that the start up effects are minimized (discussed in Chap. 7). All simulations took less than one second.

In previous chapters LC was the number of lost customers per unit time but now we will use another definition. In this chapter LC will denote the percent of arrivals that get a busy signal and hang up. Let τ be the number of customers that end up in the lost customer box in Fig. 10.1. The number of calls that end in the finished box is always 10,000. So, $LC = [\tau/(10,000 + \tau)]100$.

10.2 Case 1: First Simulation

We start with $M = 10$ and $n_x = n_y = n_z = 2$. This case is to see how the present system is performing. The fuzzy probability distributions for the system are defined in Table 10.1. Using Chap. 8 we know how to choose λ_a, λ_b, λ_c, μ_x, μ_y and μ_z to approximate the end points of α-cuts of $\overline{LC}$ and $\overline{R}$. Tables 10.2 and 10.3 has these results for the $\alpha = 0$ cut.

Table 10.1. Fuzzy Probability Distributions for Fig. 10.1

Item	Distribution	Details
Call A	Exponential	$\overline{\lambda}_a = (3/4/5)$
Call B	Exponential	$\overline{\lambda}_b = (2/3/4)$
Call C	Exponential	$\overline{\lambda}_c = (1/2/3)$
Server = X	Normal	$\overline{\mu}_x = (0.4/0.6/0.8), \sigma_x = 0.08$
Server = Y	Normal	$\overline{\mu}_y = (0.8/1.0/1.2), \sigma_y = 0.16$
Server = Z	Normal	$\overline{\mu}_z = (1.7/2.0/2.3), \sigma_z = 0.34$

Table 10.2. Alpha Zero Cut Values of the Parameters for Min LC, R

Min	λ_a	λ_b	λ_c	μ_x	μ_y	μ_z
LC	3	2	1	0.4	0.8	1.7
R	3	2	1	0.4	0.8	1.7

Table 10.3. Alpha Zero Cut Values of the Parameters for Max LC, R

Max	λ_a	λ_b	λ_c	μ_x	μ_y	μ_z
LC	5	4	3	0.8	1.2	2.3
R	5	4	3	0.8	1.2	2.3

Obvious changes to estimate the $\alpha = 0.5$ cut. Finally, we just use the $\alpha = 1$ cut values of the parameters for the alpha equal one cut of $\overline{LC}$ and $\overline{R}$. Results are in Table 10.4. A typical value for CLOCK was 1520.77 which is around three days (8 hours per day) of simulation which took less than one second.

Table 10.4. Case 1 Simulation Alpha-Cuts of $\overline{LC}$, $\overline{R}$

Item	$\alpha = 0$ Cut	$\alpha = 0.5$ Cut	$\alpha = 1$ Cut
LC	[36, 82]	[58, 77]	69
R	[3.242, 5.950]	[4.100, 5.389]	4.726

We show the graphs of $\overline{LC}$ and $\overline{R}$ in Figs. 10.2 and 10.3, respectively. Their graphs from the other cases, discussed below, are also in these figures.

10.3 Case 2: Second Simulation

The percent of lost calls is very high and management's goal is to reduce it. From the standard simulation output file we notice that the utilization of

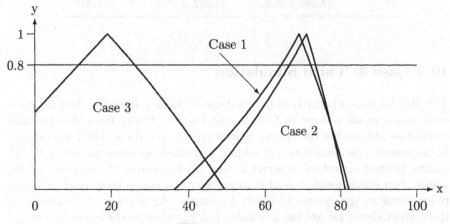

Fig. 10.2. $\overline{LC}$ (Percent) in Cases 1, 2, 3

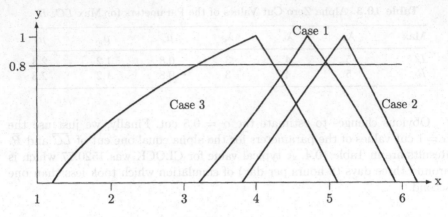

Fig. 10.3. $\overline{R}$ (Minutes) in Cases 1, 2, 3

the operators in server Z is 100%. This is one of the problems causing the high loss of calls. Server Z being too busy sends too many calls of type C to the queue filling it to capacity. Management considers a training program for the operators in server Y so they can also handle calls of type C. The only change from the system in Case 1 is that calls C first go to server Y, and if it is busy, they are sent to server Z. Also, in Table 10.1, $\mu_y = \mu_z = (1.7/2.0/2.3)$, $\sigma_y = \sigma_z = 0.34$. Simulation results for the $\alpha = 0, 0.5, 1$ cuts of $\overline{LC}$ and $\overline{R}$ are in Table 10.5. The graphs are also shown in Figs. 10.2 and 10.3.

Table 10.5. Case 2 Simulation Alpha-Cuts of $\overline{LC}$, $\overline{R}$

Item	$\alpha = 0$ Cut	$\alpha = 0.5$ Cut	$\alpha = 1$ Cut
LC	[43, 81]	[60, 77]	71
R	[3.849, 6.199]	[4.567, 5.756]	5.217

10.4 Case 3: Third Simulation

The plan for Case 2 turned out to be a disaster. In fact, from Fig. 10.2 $\overline{LC}$ even increased a small amount in Case 2 over Case 1. Again, from the standard simulation output file the operators in server Z are still at 100% utilization. Management now considers: (1) adding a trained operator to server Y; (2) adding trained operators to server Z; and (3) increasing the capacity of the queue. Top management's goal is to get lost customers to around 20% and reduce time in the system to about 4 minutes. An average of 4 minutes in the system allows for around 2 minutes holding time in the queue where the caller will hear company advertisements. Everything remains the same as in

Table 10.6. Case 3 Simulation Alpha-Cuts of $\overline{LC}$, $\overline{R}$

Item	$\alpha = 0$ Cut	$\alpha = 0.5$ Cut	$\alpha = 1$ Cut
LC	$[0, 49]$	$[0, 37]$	19
R	$[1.225, 5.078]$	$[1.863, 4.610]$	3.951

Case 2 except: (1) $n_y = 3$; (2) $n_z = 8$; and (3) $M = 20$. Simulation results are in Table 10.6 and Figs. 10.1 and 10.3. Actually, there were many other simulations before we arrived at these values for n_y, n_z and M. We have omitted all the other results. The GPSS program for this case is in Chap. 28. $\overline{LC}$ is a discrete fuzzy set whose graph is approximated by a continuous fuzzy set (Sect. 2.7).

10.5 Summary

In this chapter we studied a fuzzy system we called a "call center" shown in Fig. 10.1 with fuzzy distributions given in Table 10.1. In the initial system (called Case 1) too many calls were lost (the caller gets a busy signal and hangs up), and for those customers who do not hang up, it takes too much time to complete the call. See Figs. 10.2 and 10.3. The base of these fuzzy numbers is like a 99% confidence interval for LC or R. Management wants to reduce the percent of lost calls to around 20% and reduce time in the system to approximately 4 minutes. Case 2 is of no help. Case 3 solves the problem at a substantial cost of hiring and training 7 new operators and increasing queue size from 10 to 20. So, in practice, we would not be finished with this problem since management just added a budget constraint. Hiring 7 new operators is over budget. A new Case 4 might be to train all six of the original operators to handle all types of calls so we would have one server, with 6 operators all having their fuzzy mean service time equal to μ_z, with queue capacity still $M = 10$. Case 4 spends no new money, after training, so will come in under budget. For Case 4 where in Fig. 10.2 (10.3) will we find $\overline{LC}$ ($\overline{R}$)?

Reference

1. Getting Started SIMPROCESS, CACI Products Company, San Diego, CA, 2004.

11 Machine Shop I

11.1 Introduction

The queuing system in this chapter is shown in Fig. 11.1. This application
was adopted from a problem in ([1], p. 594). We continue this problem in the
next chapter.

We first describe the crisp queuing system in Fig. 11.1. Items arrive and
first go to the queue $Q11$ in front of machine $M11$. Of course, if $M11$ is free
the arriving unit goes directly to the machine. The production line consists
of five machines $M1i$, $i = 1, 2, 3, 4, 5$, of type $M11$ and Fig. 11.1 only shows
the first one $M11$. Each $M1i$ has a queue $Q1i$, $1 \leq i \leq 5$, in front of it. All
queues have finite capacity shown in Table 11.2.

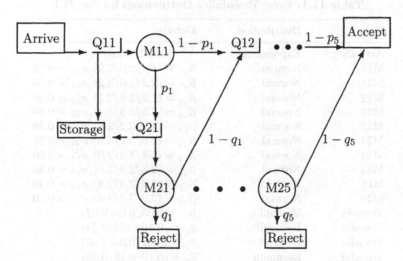

Fig. 11.1. Machine Shop I Model

Each $M1i$ may produce a defective product, $i = 1, \ldots, 5$. If the output is
determined to be defective it is sent to machine $M2i$, $1 \leq i \leq 5$, for reworking.
Each $M2i$ has a queue $Q2i$, $1 \leq i \leq 5$, of limited capacity in front of it. p_i
is the probability that the output from machine $M1i$ is defective and sent to

James J. Buckley: *Simulating Fuzzy Systems*, StudFuzz **171**, 87–93 (2005)
www.springerlink.com   © Springer-Verlag Berlin Heidelberg 2005

$M2i$, $i = 1, \ldots, 5$. So, $1 - p_i$ is the probability that the product from $M1i$ is not defective and it is sent to $Q_{1,i+1}$ for $M_{1,i+1}$, $i = 1, 2, 3, 4$. For $M15$ the probability is $1 - p_5$ that it is not defective and is finished and acceptable.

All reworking is done by machines $M2i$, $1 \leq i \leq 5$. However, some defective items cannot be salvaged. q_i is the probability that a defective unit sent to $M2i$ can not be fixed and it is sent to the "reject" box in Fig. 11.1, $i = 1, \ldots, 5$. So, $1 - q_i$ is the probability that a defective sent from $M1i$ to $M2i$ has been fixed and then sent to $Q_{1,i+1}$ for $M_{1,i+1}$, $i = 1, 2, 3, 4$. For $M25$ the probability is $1 - q_5$ it is sent to "finished-accepted".

Once the production line is set up and running it continues for 24 hours, seven days per week, until the total production quota is met. What happens if queue capacity is exceeded? This is the "storage" box in Fig. 11.1. Since production can not be disturbed once begun, all excess queue capacity is sent to a storage facility. At a later time, when we have some slack time in production, or using overtime after the production run is finished, the items in storage are put into the system to be finished.

Arrival times will be modelled using the exponential and all service times are modelled by the normal distribution. The parameters in these distributions all need to be estimated from data and their fuzzy estimators are given in Table 11.1. Also, the probabilities p_i and q_i, $1 \leq i \leq 5$, are estimated and

Table 11.1. Fuzzy Probability Distributions for Fig. 11.1

Item	Distribution	Details
Arrivals	Exponential	$\overline{\lambda} = (0.4/0.5/0.6)$
M11	Normal	$\overline{\mu}_{11} = (1.8/2.0/2.2), \sigma_{11} = 0.36$
M21	Normal	$\overline{\mu}_{21} = (2.8/3.0/3.2), \sigma_{21} = 0.56$
M12	Normal	$\overline{\mu}_{12} = (1.5/1.8/2.1), \sigma_{12} = 0.30$
M22	Normal	$\overline{\mu}_{22} = (2.5/2.8/3.1), \sigma_{22} = 0.50$
M13	Normal	$\overline{\mu}_{13} = (1.9/2.2/2.5), \sigma_{13} = 0.38$
M23	Normal	$\overline{\mu}_{23} = (2.9/3.2/3.5), \sigma_{23} = 0.58$
M14	Normal	$\overline{\mu}_{14} = (1.8/1.9/2.0), \sigma_{14} = 0.36$
M24	Normal	$\overline{\mu}_{24} = (2.8/2.9/3.0), \sigma_{24} = 0.56$
M15	Normal	$\overline{\mu}_{15} = (2.0/2.4/2.8), \sigma_{15} = 0.40$
M25	Normal	$\overline{\mu}_{25} = (3.0/3.4/3.8), \sigma_{25} = 0.60$
Transfer	Bernoulli	$\overline{p}_1 = (0.03/0.05/0.07)$
Transfer	Bernoulli	$\overline{q}_1 = (0.16/0.20/0.24)$
Transfer	Bernoulli	$\overline{p}_2 = (0.02/0.03/0.04)$
Transfer	Bernoulli	$\overline{q}_2 = (0.10/0.15/0.20)$
Transfer	Bernoulli	$\overline{p}_3 = (0.02/0.04/0.06)$
Transfer	Bernoulli	$\overline{q}_3 = (0.14/0.18/0.22)$
Transfer	Bernoulli	$\overline{p}_4 = (0.05/0.06/0.07)$
Transfer	Bernoulli	$\overline{q}_4 = (0.20/0.25/0.30)$
Transfer	Bernoulli	$\overline{p}_5 = (0.06/0.08/0.10)$
Transfer	Bernoulli	$\overline{q}_5 = (0.15/0.20/0.25)$

Table 11.2. Initial Queue Capacities in Fig. 11.1

Queue	Capacity
Q11	50
Q21	20
Q12	50
Q22	20
Q13	50
Q23	20
Q14	50
Q24	20
Q15	50
Q25	20

their fuzzy estimators are also in Table 11.1. All the fuzzy estimators are triangular fuzzy numbers. Time units are in minutes.

The plant manager is interested in first obtaining for the initial system the fuzzy numbers for response time $\overline{R}$, throughput $\overline{X}$ and storage $\overline{S}$. $\overline{R}$ is the average time a unit spends in the system and $\overline{X}$ is the average number of finished and accepted units per 24 hours. $\overline{S}$, new to this chapter, will be the percent of all items arriving at $Q11$ that end up in storage.

We should also determine the percent of all units arriving at $Q11$ that get rejected. All arrivals either go to storage, get rejected or are finished and accepted. Then management's goal is to maximize $\overline{X}$ and minimize $\overline{R}$ and $\overline{S}$. We have no one-step formulas for R, X and S so we use simulation to estimate their $\alpha = 0, 0.5, 1$ cuts (see Chap. 7). All simulations will run until 50,000 items end up in the accepted box. The value of 50,000 should be sufficiently larger to put the system well into steady-state and make the initial conditions negligible (swamping method, Chap. 7).

Let n_s be the number of items that go to storage and let n_r be all the units that end up rejected. Then $S = (n_s/(50,000 + n_s + n_r)$. If CLOCK is the total minutes of simulation till 50,000 get accepted, then $X = (50,000/CLOCK) \bullet (60)(24)$. Now R is the mean time in the system for all (storage, rejected, accepted) units, excluding those that spend zero time in the system, so the average time in the system for an accepted item will be slightly higher than the values reported below. The units that spend zero time in the system are those who go directly to storage at arrival because Q11 is full (at capacity).

11.2 Case 1: First Simulation

This case is to see how the present system is performing. The fuzzy probability distributions for the system are defined in Table 11.1. Note that the numbers for $\overline{\lambda}$ are given as rates (number of arrivals per time unit) so the reciprocal is the average time between arrivals. The values for $\overline{\mu}$ are the mean service times. Recall (Chap. 8) that we will use crisp estimators for the standard deviation of the normal distribution. Using Chap. 8 we know how to choose λ, μ_{11}, $\mu_{21}, \ldots, \mu_{25}$ to approximate the end points of the α-cuts of $\overline{R}$ and $\overline{X}$. We need to determine this for $\overline{S}$ also, and how to pick the values for the p_i and q_i, $1 \le i \le 5$. Let $\overline{S}[\alpha] = [s_1(\alpha), s_2(\alpha)]$. We found that $min(\lambda)$, $min(\mu)_{ij}$ gives $s_1(\alpha)$ and $max(\lambda)$, $max(\mu)_{ij}$ is for $s_2(\alpha)$. We will have to experiment with the p_i and q_i values, because they will depend on $\overline{\lambda}$ and the $\overline{\mu}_{ij}$, to approximate the end points of the alpha-cuts. We ran simulations using max/min values of λ, μ_{ij}, p_i and the q_i to see what combinations produce the max/min of R, X and S. These results are in Tables 11.3 and 11.4. In the tables: (1) "min" means use the left end point of the interval for the α-cut of the fuzzy estimator; and (2) "max" means use the right end point. In Table 11.3 "max/min" means both values give the same result. The information in Tables 11.3 and 11.4 can now be used to estimate alpha-cuts of $\overline{R}$, $\overline{X}$ and $\overline{S}$ with the simulation results in Table 11.5. If there are alternate optimal solutions for the values of the parameters we only list one solution.

The value of CLOCK and the run time varies with the values of the parameters. However, the simulation is usually between 90 and 100 days (24 hours per day of production) and the run time varied from 2 seconds to 13 seconds.

Table 11.3. Values of the Parameters for Min R, X, S

Min	λ	μ_{ij}	p_i	q_i
R	min	min	max	max
X	min	max	max	max
S	min	min	max/min	max/min

Table 11.4. Values of the Parameters for Max R, X, S

Max	λ	μ_{ij}	p_i	q_i
R	max	max	min	min
X	max	min	min	min
S	max	max	min	min

Table 11.5. Case 1 Simulation Alpha-Cuts of $\overline{R}$, $\overline{X}$, $\overline{S}$

Item	$\alpha = 0$ Cut	$\alpha = 0.5$ Cut	$\alpha = 1$ Cut
$\overline{R}$	$[15, 334]$	$[32, 314]$	250
$\overline{X}$	$[501, 727]$	$[542, 647]$	591
$\overline{S}$	$[0.00, 39.1]$	$[0.00, 27.7]$	12.9

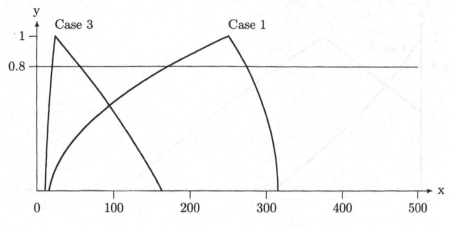

Fig. 11.2. $\overline{R}$ (Minutes) in Cases 1 and 3

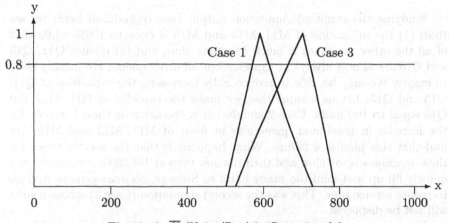

Fig. 11.3. $\overline{X}$ (Units/Day) in Cases 1 and 3

We show the graphs of $\overline{R}$, $\overline{X}$ and $\overline{S}$ in Figs. 11.2, 11.3 and 11.4, respectively. Their graphs from the other cases, discussed below, are also in these figures.

11.3 Cases 2 and 3: Second and Third Simulation

In Fig. 11.4 it is the right end point of the base of $\overline{S}$ that worries the plant manager. The base of the fuzzy number is like a 99% confidence interval. The maximum value of S for $\alpha = 0$ was 39.1% which comes from 33,140 units to storage, 1,668 rejected and 50,000 finished and accepted. Having 33,140 in storage is impossible. It can not be done. The manager decides to temporarily forget the other goals ($min\overline{R}$, $max\overline{X}$) and concentrate on reducing $\overline{S}$.

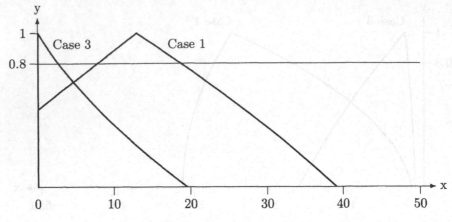

Fig. 11.4. $\overline{S}$ (Percent) to Storage in Cases 1 and 3

Studying the standard simulation output (not reproduced here) we see that: (1) the utilization of M11, M13 and M15 is close to 100% while that of all the other machines is much less than that; and (2) queues Q11, Q13 and Q15 are almost always at capacity but all other queues are usually close to empty. We may be able to reduce $\overline{S}$ by increasing the capacities of Q11, Q13 and Q15. Let us assume that we make the capacity of Q11, Q13 and Q15 equal to 100 units. Case 2 simulation is the same as Case 1 except for the increase in maximum queue size in front of M11, M13 and M15. We find that this plan is a failure. What happens is that the service times for these machines is too slow and their queues, even at $100, 200, \ldots$ capacity, will quickly fill up and send too many units to Storage. So management decides to reduce service time. This was the second simulation (Case 2) whose results will not be displayed.

Now we move on to the third simulation (Case 3). Management decides to spend the money to upgrade machines $M11$, $M13$ and $M15$ to new machines with expected shorter service time. The new service times are estimated to be $\overline{\mu}_{11} = \overline{\mu}_{13} = \overline{\mu}_{15} = (1.2/1.5/1.8)$ and $\sigma_{11} = \sigma_{13} = \sigma_{15} = 0.24$. The third simulation is now the same as Case 1, same queue capacities as in Table 11.2, with different fuzzy means and crisp standard deviations as defined above. The results are in Table 11.6 with graphs in Figs. 11.2–11.4. Notice that in this chapter $\overline{R}$ and $\overline{X}$ are discrete fuzzy sets whose graphs were approximated by a continuous fuzzy set (Sect. 2.7).

11.4 Summary

The plant manager is pleased with the results of Case 3: (1) $\overline{R}$ = response time has been substantially reduced; (2) $\overline{X}$ = throughput has been significantly

Table 11.6. Case 3 Simulation Alpha-Cuts of $\overline{R}$, $\overline{X}$, $\overline{S}$

Item	$\alpha = 0$ Cut	$\alpha = 0.5$ Cut	$\alpha = 1$ Cut
$\overline{R}$	$[10, 163]$	$[13, 107]$	23
$\overline{X}$	$[525, 785]$	$[605, 752]$	702
$\overline{S}$	$[0, 19.5]$	$[0, 6.0]$	0

increased; and (3) $\overline{S}$ = percent into storage has also been greatly decreased. However, $max\overline{S}[0] = 19.5\%$ is still high. So, before management spends all that money to obtain new machines they want to explore other alternatives. One alternative suggested by the engineers is to put the production line into another configuration (rearrange the machines). This will be the topic of the next chapter.

Reference

1. H.A. Taha: Operations Research, Fifth Edition, Macmillan, N.Y., 1992.

12 Machine Shop II

12.1 Introduction

This chapter continues the machine shop model discussed in the previous chapter. Management is considering a radical change in the machine shop discussed in Chap. 11. The new setup is shown in Fig. 12.1. All the machines M21, M22, M23, M24, M25 will be put into one group of three machines M31, M32 and M33. Service station M3 will consist of the three parallel and identical servers M31, M32, M33. The workers in M3 will be trained to service units from all the five machines M11, ..., M15. This will produce a savings of two machines and this savings will be put into new machines M11a for M11, M13a for M13 and M15a for M15 all having faster service times.

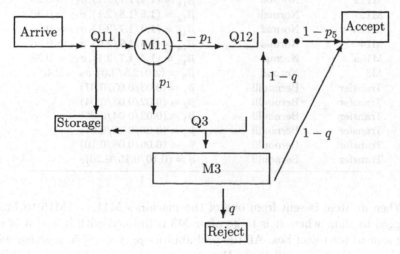

Fig. 12.1. Machine Shop II Model

The new service times for M31, M32, M33 and M11a, M13a, M15a are all estimated by "experts". Since we do not have any data on these new machines we ask certain experts to estimate their service time. That is, we have no statistical data to generate the fuzzy numbers from a set of

James J. Buckley: *Simulating Fuzzy Systems*, StudFuzz **171**, 95–101 (2005)
www.springerlink.com   © Springer-Verlag Berlin Heidelberg 2005

confidence intervals as discussed in Chap. 3. So let us briefly see how this may be accomplished (also discussed in Sect. 3.2 in Chap. 3). First assume we have only one expert and he/she is to estimate the value of some service rate μ. We can solicit this estimate from the expert as is done in estimating job times in project scheduling ([1], Chap. 13). Let $a =$ the "pessimistic" value of μ, or the smallest possible value, let $c =$ be the "optimistic" value of μ, or the highest possible value, and let $b =$ the most likely value of μ. We then ask the expert to give values for a, b, c and we construct the triangular fuzzy number $\overline{\mu} = (a/b/c)$ for μ. If we have a group of N experts all to estimate the value of μ we solicit the a_i, b_i and c_i, $1 \leq i \leq N$, from them. Let a be the average of the a_i, b is the mean of the b_i and c is the average of the c_i. The simplest thing to do is to use $(a/b/c)$ for $\overline{\mu}$. Our fuzzy estimators are given in Table 12.1. The values for $\overline{\mu}_{11}$, $\overline{\mu}_{13}$ and $\overline{\mu}_{15}$ in Table 12.1 are different from the values used in Case 3 in Chap. 11 because these are different new machines. The three machines M31, M32, M33 in M3 all have the same $\overline{\mu}_3$ and σ_3 given in Table 12.1.

Table 12.1. Fuzzy Probability Distributions for Fig. 12.1

Item	Distribution	Details
Arrivals	Exponential	$\overline{\lambda} = (0.4/0.5/0.6)$
M11a	Normal	$\overline{\mu}_{11} = (1.4/1.7/2.1), \sigma_{11} = 0.28$
M12	Normal	$\overline{\mu}_{12} = (1.5/1.8/2.1), \sigma_{12} = 0.30$
M13a	Normal	$\overline{\mu}_{13} = (1.4/1.7/2.1), \sigma_{13} = 0.28$
M14	Normal	$\overline{\mu}_{14} = (1.8/1.9/2.0), \sigma_{14} = 0.36$
M15a	Normal	$\overline{\mu}_{15} = (1.4/1.7/2.1), \sigma_{15} = 0.28$
M3	Normal	$\overline{\mu}_{3} = (2.0/2.5/3.0), \sigma_{3} = 0.40$
Transfer	Bernoulli	$\overline{p}_1 = (0.03/0.05/0.07)$
Transfer	Bernoulli	$\overline{p}_2 = (0.02/0.03/0.04)$
Transfer	Bernoulli	$\overline{p}_3 = (0.02/0.04/0.06)$
Transfer	Bernoulli	$\overline{p}_4 = (0.05/0.06/0.07)$
Transfer	Bernoulli	$\overline{p}_5 = (0.06/0.08/0.10)$
Transfer	Bernoulli	$\overline{q} = (0.10/0.15/0.20)$

When an item is sent from one of the machines M11,..., M15 to M3 it is tagged to show where it is to go after M3 is finished with it and it is not being sent to the reject box. All the probabilities p_i, $1 \leq i \leq 5$, together with their fuzzy estimators, still apply. However, now there is only one probability attached to M3 which is $q =$ the probability it is rejected so that $1 - q$ is the probability it gets sent back into the major production line. The fuzzy estimator for q, obtained from expert opinion, is in Table 12.1.

The goal is still to $min\overline{R}$, $max\overline{X}$ and $min\overline{S}$. However, we first need to simulate the proposed new system and estimate these fuzzy numbers to see

Table 12.2. Initial Queue Capacities in Fig. 12.1

Queue	Capacity
Q11	50
Q12	50
Q13	50
Q14	50
Q15	50
Q3	50

how it is operating. All simulations will run until 50,000 units are accepted. The values of R, X and S are computed as in the previous chapter.

12.2 Case 1: First Simulation

This case is to see how the proposed system is performing. The fuzzy probability distributions for the system are defined in Table 12.1. Note again that the numbers for $\overline{\lambda}$ are given as rates (number of arrivals per time unit) so the reciprocal is the average time between arrivals. The values for $\overline{\mu}$ are the mean service times. Using Chaps. 8 and 11 we know how to choose λ, $\mu_{11}, \ldots, \mu_{15}$ and μ_3 to approximate the end points of the α-cuts of $\overline{R}$, $\overline{X}$ and $\overline{S}$. We will have to experiment again with the p_i and q values, because they will depend on $\overline{\lambda}$ and the $\overline{\mu}_{ij}$, to approximate the end points of the alpha-cuts. We ran simulations using max/min values of λ, μ_{ij}, p_i and q to see what combinations produce the max/min of R, X and S. These results are in Tables 12.3 and 12.4. In the tables: (1) "min" means use the left end point of the interval for the α-cut of the fuzzy estimator; and (2) "max" means use the right end point. In Table 12.3 "max/min" means both values give the same

Table 12.3. Values of the Parameters for Min R, X, S

Min	λ	μ_{ij}	p_i	q
R	min	min	min	max
X	min	max	max	max
S	min	min	max/min	max/min

Table 12.4. Values of the Parameters for Max R, X, S

Max	λ	μ_{ij}	p_i	q
R	max	max	max	min
X	max	min	min	min
S	max	max	max	min

Table 12.5. Case 1 Simulation Alpha-Cuts of $\overline{R}, \overline{X}, \overline{S}$

Item	$\alpha = 0$ Cut	$\alpha = 0.5$ Cut	$\alpha = 1$ Cut
$\overline{R}$	[12.5, 310.9]	[25.2, 189.9]	175
$\overline{X}$	[528, 714]	[600, 683]	654
$\overline{S}$	[0.00, 34.74]	[0.00, 19.32]	4.8

result. The information in Tables 12.3 and 12.4 can now be used to estimate alpha-cuts of $\overline{R}, \overline{X}$ and $\overline{S}$ with the simulation results in Table 12.5.

In a simulation the value of CLOCK and the run time varies with the values of the parameters. Time units are in minutes. However, the simulation is usually between 90 and 100 days (24 hours per day of production) and the run time varied from 2 seconds to 13 seconds.

We show the graphs of $\overline{R}, \overline{X}$ and $\overline{S}$ in Figs. 12.2, 12.3 and 12.4, respectively. Their graphs from the other cases, discussed below, are also in these figures. A sample GPSS program used in this chapter is in Chap. 28.

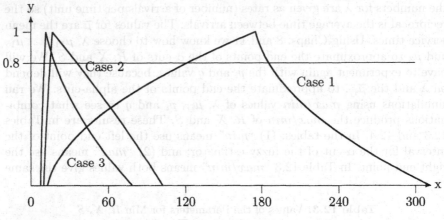

Fig. 12.2. $\overline{R}$ (Minutes) in Cases 1, 2, 3

12.3 Case 2: Second Simulation

Management is really worried about the results on storage. The base of $\overline{S}$ is like a 99% confidence interval and the right end point of about 35% corresponding to 27,709 units sent to storage. This is not possible since there is no room to hold that many items. As the number of items in storage grows production would need to be halted to process these storage units and the plant manager does not wish to stop the production line.

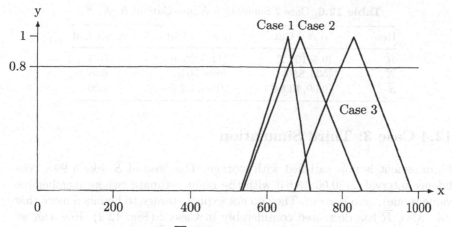

Fig. 12.3. $\overline{X}$ (Units/Day) in Cases 1, 2, 3

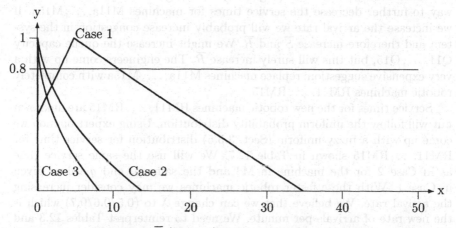

Fig. 12.4. $\overline{S}$ (Percent) to Storage in Cases 1, 2, 3

We notice from the standard simulation output that the machines in M3 are not being fully utilized. So let us consider deleting two machines from M3. The savings from this change will be put into decreasing the service times on machines M11a, ..., M15a. The engineers think they can improve the service times slightly to: $\overline{\mu}_{11} = (1.3/1.4/1.5)$, $\overline{\mu}_{12} = (1.4/1.5/1.6)$, $\overline{\mu}_{13} = (1.3/1.4/1.5)$, $\overline{\mu}_{14} = (1.4/1.5/1.6)$, $\overline{\mu}_{15} = (1.3/1.4/1.5)$. The corresponding standard deviations are $\sigma_{11} = 0.26$, $\sigma_{12} = 0.28$, $\sigma_{13} = 0.26$, $\sigma_{14} = 0.28$, $\sigma_{15} = 0.26$. These will be the only changes from Case 1. Now simulate the system to estimate $\overline{S}$ and $\overline{R}$, $\overline{X}$. The results are in Table 12.6 and Figs. 12.2–12.4.

Table 12.6. Case 2 Simulation Alpha-Cuts of $\overline{R}$, $\overline{X}$, $\overline{S}$

Item	$\alpha = 0$ Cut	$\alpha = 0.5$ Cut	$\alpha = 1$ Cut
$\overline{R}$	[9.5, 101.2]	[11.5, 65.4]	16.4
$\overline{X}$	[534, 850]	[608, 764]	686
$\overline{S}$	[0.00, 13.64]	[0.00, 1.25]	0.00

12.4 Case 3: Third Simulation

Management is now satisfied with storage. The base of $\overline{S}$, like a 99% confidence interval, is $[0.00, 13.64]$ with the point estimate (where membership value is one), zero percent. They do not expect storage to be much more that 0%. Also, $\overline{R}$ has decreased considerably in Case 2 (Fig. 12.2). How can we increase throughput $\overline{X}$ without increasing $\overline{S}$ and $\overline{R}$? Notice that Case 2 did not substantially increase $\overline{X}$ (Fig. 12.3). The engineers say that there is no way to further decrease the service times for machines M11a, . . . , M15a. If we increase the arrival rate we will probably increase congestion in the system and therefore increase $\overline{S}$ and $\overline{R}$. We might increase the queue capacity Q11, . . . , Q15, but this will surely increase $\overline{R}$. The engineers come up with a very expensive suggestion: replace machines M11a, . . . , M15a with completely robotic machines RM11, . . . , RM15.

Service times for the new robotic machines RM11, . . . , RM15 are unknown but will follow the uniform probability distribution. Using expert opinion we come up with a fuzzy uniform (Sect. 3.5.5) distribution for service time for RM11, . . . , RM15 shown in Table 12.7. We will use the same service time as in Case 2 for the machine in M3 and the same p_i and q values given in Case 1. With these faster robotic machines we may consider increasing the arrival rate. We believe that we can change $\overline{\lambda}$ to $(0.5/0.6/0.7)$ which is the new rate of arrivals per minute. We need to reinterpret Tables 12.3 and 12.4 for the fuzzy uniform distribution. Min under μ_{ij} means use the left end points of $\overline{a}_{1j}[0]$ and $\overline{b}_{1j}[0]$ and max under μ_{ij} means use the right end points. Queue capacities are the same as in Table 12.2. Now we are ready to simulate the fuzzy system to estimate $\overline{X}$ and $\overline{R}$, $\overline{S}$. The results are in

Table 12.7. Fuzzy Uniform Probability Distributions for Case 3

Item	Distribution	$\overline{a}$	$\overline{b}$
RM11	Uniform	$\overline{a}_{11} = (0.9/1.0/1.1)$	$\overline{b}_{11} = (1.1/1.2/1.3)$
RM12	Uniform	$\overline{a}_{12} = (1.0/1.1/1.2)$	$\overline{b}_{12} = (1.2/1.3/1.4)$
RM13	Uniform	$\overline{a}_{13} = (0.9/1.0/1.1)$	$\overline{b}_{13} = (1.1/1.2/1.3)$
RM14	Uniform	$\overline{a}_{14} = (1.0/1.1/1.2)$	$\overline{b}_{14} = (1.2/1.3/1.4)$
RM15	Uniform	$\overline{a}_{15} = (0.9/1.0/1.1)$	$\overline{b}_{15} = (1.1/1.2/1.3)$

Table 12.8. Case 3 Simulation Alpha-Cuts of $\overline{R}$, $\overline{X}$, $\overline{S}$

Item	$\alpha = 0$ Cut	$\alpha = 0.5$ Cut	$\alpha = 1$ Cut
$\overline{R}$	$[7.1, 77.8]$	$[7.7, 24.8]$	11.3
$\overline{X}$	$[671, 980]$	$[738, 906]$	829
$\overline{S}$	$[0.00, 8.08]$	$[0.00, 0.02]$	0.00

Table 12.8 and Figs. 12.2–12.4. In this chapter $\overline{X}$ is a discrete fuzzy set whose graph is approximated by a continuous fuzzy set (Sect. 2.7).

12.5 Summary

Let $\overline{R}_i$, $\overline{X}_i$ and $\overline{S}_i$ be the results from Cases $i = 1, 2, 3$. From our method of ranking fuzzy numbers in Sect. 2.5 we see from Figs. 12.2–12.4 that: (1) $\overline{R}_2 \approx \overline{R}_3 < \overline{R}_1$; (2) $\overline{X}_1 \approx \overline{X}_2 < \overline{X}_3$; and (3) $\overline{S}_2 \approx \overline{S}_3 < \overline{S}_1$. The bases of these fuzzy numbers are like 99% confidence intervals. Clearly, management prefers Case 3 to Cases 1 and 2.

Now we compare these results to those in Chap. 11. We compare: (1) Fig. 11.2 and Fig. 12.2 for $\overline{R}$; (2) Fig. 11.3 and Fig. 12.3 for $\overline{X}$; and (3) Fig. 11.4 and Fig. 12.4 for $\overline{S}$. The result is that Case 3 of Chap. 12 is the best. So, the plant manager now needs to go to corporate headquarters to secure the funds needed to implement Case 3.

Reference

1. H.A. Taha: Operations Research, Fifth Edition, Macmillan, N.Y., 1992.

13 Emergency Room Model

13.1 Introduction

The system in this chapter was adapted from an example in [1]. The flow through the system is shown in Fig. 13.1. Patients arrive, according to the exponential distribution, by coming through the front door or by ambulance. We model this emergency room during its busiest eight hour shift. All patients are classified as A or B. All ambulance patients are classified as A patients. When a patient comes through the front door a nurse (N in Fig. 13.1) immediately assigns them as a class A, or B, patient. We assume there is no significant time delay in this initial classification.

Class B patients first go to the "sign in" procedure (S in Fig. 13.1) where a secretary gets their basic information. After the sign in class B patients proceed to an evaluation station manned by nurses (E in Fig. 13.1). We assume that all service stations have unlimited waiting room preceding them. This emergency room will not turn any patients away because of limited queue capacity. When it gets busy patients can sit on the floor, wait in the halls or even wait outside. Because all queues have infinite capacity they were not shown in Fig. 13.1. Also, all service times will be modelled by the normal distribution.

All patients will go to T1 in Fig. 13.1. T1 is where the patient is transported to a "room" for treatment. There are n treatment rooms and this transfer is handled by an emergency room assistant. Right now there are $n = 6$ treatment rooms and hospital management is not considering adding more rooms. Once inside the treatment area class B patients go through "paper work", P1 in Fig. 13.1, which is done by nurses. All patients are then treated by doctors and nurses (T2 in Fig. 13.1). After treatment class B patients are released or admitted to the hospital, but class A patients now have the "paper work" done (P2 in Fig. 13.1), which is handled by nurses, before release/admit. The operation of release/admit is assumed to have no time delay.

Hospital management wants to "optimize" their emergency room. They have too many complaints about their emergency room takes too much time. Their goals are to minimize R = mean time spent in the emergency room and minimize costs. Since we do not have any one-step formulas for R in this model we will simulate it to estimate alpha-cuts of $\overline{R}$ (see Chap. 7). The

James J. Buckley: *Simulating Fuzzy Systems*, StudFuzz **171**, 103–109 (2005)
www.springerlink.com

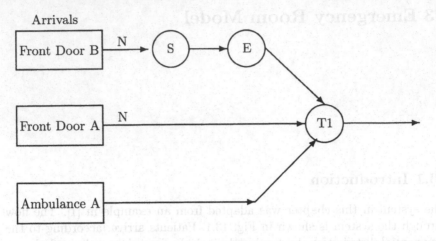

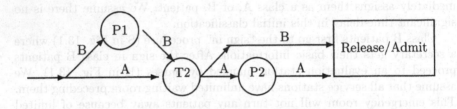

Fig. 13.1. Emergency Room

costs associated with this system will be the personnel costs for one eight hour shift. This cost will be crisp and not fuzzy. Personnel costs are made up from the secretaries, nurses, emergency room assistants and doctors. These hourly costs, including fringe benefit expenses, etc. are in Table 12.1.

These two goals will be in conflict. To reduce $\overline{R}$ we add more personnel which increases the cost. Management decides that reducing $\overline{R}$ is twice as important than minimizing costs. So our objective now is to

$$min[\overline{Z} = \tau_1 \overline{R} + \tau_2 Costs], \tag{13.1}$$

for $\tau_i > 0$ and $\tau_1 + \tau_2 = 1$, with special interest for $\tau_1 = 2/3$, $\tau_1 = 1/3$. Now find the personnel assignments to minimize $\overline{Z}$ for selected values of the τ_i.

All simulations will attempt to run until 50,000 patients are in the box release/admit. We need to run the simulation a long time so as not to bias R with the start-up results of all servers empty (the swamping method in Chap. 7).

13.2 Case 1: First Simulation

This case is to see how the present system is performing. From the data collected on the emergency room we estimate the arrival rates and the service times. As in Chap. 3 these will be fuzzy estimators. We discussed in Chap. 8 that we will use only a point estimator for the standard deviation of the fuzzy normal distribution. The fuzzy probability distributions for the system are defined in Table 13.2. Note that the numbers for $\overline{\lambda}$ are given as rates (number of arrivals per time unit) so the reciprocal is the average time between arrivals. The values for $\overline{\mu}$ are the mean service times. Using Chap. 8 we know how to choose the λ and the μ to approximate the end points of the α-cuts of $\overline{R}$. In general we use: (1) $min\lambda$ and $min\mu$ for $minR$; but (2) $max\lambda$ and $max\mu$ for $maxR$. The simulation results are in Table 13.3.

The current personnel costs from Table 13.1 is \$940 per hour. Putting Table 13.3 together with the \$940 we may compute $\overline{Z}$, at least its $\alpha = 0, 0.5, 1$ cuts, from (13.1) and construct the graphs in Figs. 13.3–13.5. We did not compute the right end points of $\overline{R}[0]$ and $\overline{R}[0.5]$ because of memory constraints (ran out of memory, see Chap. 6). However, since $\overline{R}[1]$ was so large we will not need the right side of $\overline{R}$.

Table 13.1. Personnel Costs in the Emergency Room

Station	Personnel	Number	Cost (\$/hour)
N	Nurse	1	65
S	Secretary	1	35
E	Nurse	2	65
T1	Assistant	2	30
P1	Secretary	1	35
T2	Nurse	2	65
	Doctor	3	150
P2	Secretary	1	35

Table 13.2. Fuzzy Probability Distributions for Fig. 13.1

Item	Distribution	Details
Front Door B	Exponential	$\overline{\lambda}_b = (0.08/0.1/0.12)$
Front Door A	Exponential	$\overline{\lambda}_{a1} = (0.02/0.03/0.04)$
Ambulance A	Exponential	$\overline{\lambda}_{a2} = (0.02/0.03/0.04)$
S	Normal	$\overline{\mu}_s = (6/8/10), \sigma_s = 1.2$
E	Normal	$\overline{\mu}_e = (8/10/12), \sigma_e = 1.6$
T1	Normal	$\overline{\mu}_{t1} = (4/5/6), \sigma_{t1} = 0.8$
P1	Normal	$\overline{\mu}_{p1} = (9/10/11), \sigma_{p1} = 1.8$
T2	Normal	$\overline{\mu}_{t2} = (25/30/35), \sigma_{t2} = 5.0$
P2	Normal	$\overline{\mu}_{p2} = (9/10/11), \sigma_{p2} = 1.8$

Table 13.3. Case 1 Simulation Alpha-Cuts of $\overline{R}$

Item	$\alpha = 0$ Cut	$\alpha = 0.5$ Cut	$\alpha = 1$ Cut
$\overline{R}$	$[56.7, *]$	$[73.5, *]$	401.6

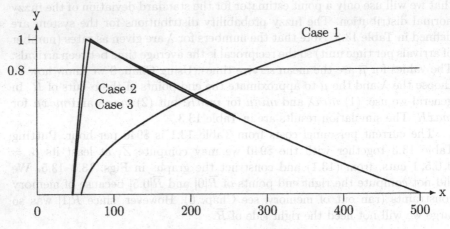

Fig. 13.2. $\overline{R}$ (Minutes) in Cases 1, 2, 3

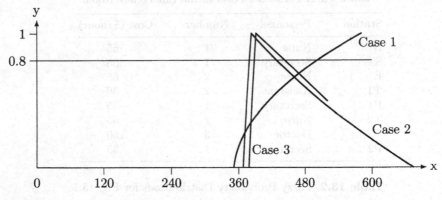

Fig. 13.3. $\overline{Z}$, $\tau_1 = 2/3$, $\tau_2 = 1/3$, in Cases 1, 2, 3

The value of CLOCK and the run time varies with the values of the parameters. The time units are in minutes. However, the simulation is usually around 650 eight hour shifts and computing times about 2 seconds. We show the graph of $\overline{R}$ in Fig. 13.2 and $\overline{Z}$ in Figs. 13.3–13.5. Their graphs from the other cases, discussed below, are also in these figures. One of the GPSS programs used in this chapter is in Chap. 28.

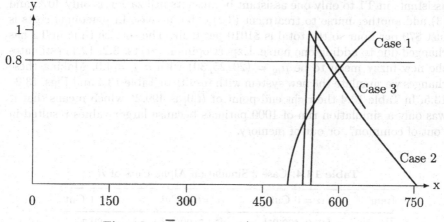

Fig. 13.4. $\overline{Z}$, $\tau_1 = \tau_2 = 1/2$, in Cases 1, 2, 3

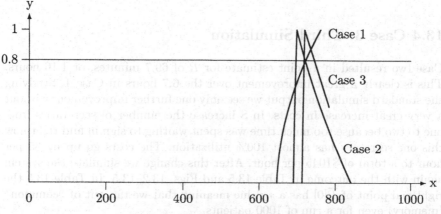

Fig. 13.5. $\overline{Z}$, $\tau_1 = 1/3$, $\tau_2 = 2/3$, in Cases 1, 2, 3

13.3 Case 2: Second Simulation

The point estimate for time spent in the system, $\overline{R}[1]$, is 401.6 minutes, or 6.7 hours. Recall that this is for the "busy" time at the emergency room. No wonder hospital management has found people sleeping in the emergency room waiting area. Each patient usually arrives with some family members which produces a lot of people waiting an average of 6.7 hours. No wonder there have been many complaints.

After studying the standard simulation output file management decides to try the following changes: (1) add one secretary to P1 because it now has almost 100% utilization and the mean time spent at this service station is too long (333 minutes) all in waiting to do the paper work; (2) reduce the two assistants in T1 to only one assistant because its utilization is only 40%; and (3) add another nurse to treatment (T2). The increase in personnel costs is just $70 per hour so the total is $1010 per hour. The service time in T2 has changed due to adding one nurse. Expert opinion (Sects. 3.2, 12.1) estimates the new fuzzy mean to be $\overline{\mu}_{t2} = (20/25/30)$ with $\sigma_{t2} = 4.0$. Making these changes we simulate the new system with results in Table 13.4 and Figs. 13.2–13.5. In Table 13.4 the right end point of $\overline{R}[0]$ is 499.2* which means that it was only a simulation run of 1000 patients because larger values resulted in "out of common", or out of memory.

Table 13.4. Case 2 Simulation Alpha-Cuts of $\overline{R}$

Item	$\alpha = 0$ Cut	$\alpha = 0.5$ Cut	$\alpha = 1$ Cut
$\overline{R}$	$[46.2, 499.2^*]$	$[54.5, 226.4]$	69.7

13.4 Case 3: Third Simulation

Case two resulted in a point estimate for R of 69.7 minutes, or 1.16 hours. This is clearly a great improvement over the 6.7 hours in Case 1. Studying the standard simulation output we see only one further improvement without a very great increase in costs. In S increase the number of secretaries from one to two because too much time was spent waiting to sign in and right now this one secretary has almost 100% utilization. The costs go up by 35 per hour to a total of $1045 per hour. After this change we simulate the system again with the outcome in Table 13.5 and Figs. 13.2–13.5. In Table 13.5 the right end point of $\overline{R}[0]$ has a * value meaning that we ran out of "common" (memory) even for a run of 1000 patients.

Table 13.5. Case 3 Simulation Alpha-Cuts of $\overline{R}$

Item	$\alpha = 0$ Cut	$\alpha = 0.5$ Cut	$\alpha = 1$ Cut
$\overline{R}$	$[45.3, *]$	$[52.5, 257.5]$	63.6

13.5 Summary

Let us look more closely at Case 3. Patient type B goes through S, E, T1, P1 and T2 in Fig. 13.1. Add up the mean service times for these stations and we get 58 minutes. But $\overline{R}[1] = 63.6$ minutes only 5.6 minutes more than the 58. Patient type A goes through T1, T2 and P2 in Fig. 13.1 with mean service times adding up to 40 minutes 23.6 under the 63.6. Hospital management likes these results. However, there is still a possibility of long waits since the right end point of $\overline{R}[0]$ is large (probably more than 5 hours).

Next we study Figs. 13.2–13.5. Let $\overline{R}_i$ ($\overline{Z}_i$) be the value of $\overline{R}$ ($\overline{Z}$) in Case $i = 1, 2, 3$. The bases of all these fuzzy numbers are like 99% confidence intervals. The reader may need to review Sect. 2.5 of Chap. 2 on how we will order (rank) fuzzy numbers. First, from Fig. 13.2, we see that $\overline{R}_2 \approx \overline{R}_3 < \overline{R}_1$, which is no surprise. The next three figures show how we may treat the situation of having two goals of $min\overline{R}$ and $min(Costs)$. Figure 13.3 shows where we weight $\overline{R}$ twice as much as "costs" and $\overline{Z}_2 \approx \overline{Z}_3 < \overline{Z}_1$. In fact, $\overline{Z}_2$ and $\overline{Z}_3$ are almost identical. Next we weight $\overline{R}$ and "costs" equally and $\overline{Z}_3$ has moved a little to the right of $\overline{Z}_2$ but still $\overline{Z}_2 \approx \overline{Z}_3 < \overline{Z}_1$. The last case is were we weight "costs" twice as much as $\overline{R}$ and Fig. 13.5 shows $\overline{Z}_1 \approx \overline{Z}_2 \approx \overline{Z}_3$. Hospital management sees no clear choice in Cases 1, 2, 3 so they request further possible changes in the system and more simulations.

Reference

1. Getting Started SIMPROCESS, CACI Products Company, San Diego, CA, 2004.

14 Machine Servicing Problem

14.1 Introduction

The machine servicing problem is shown in Fig. 14.1 and a basic description can be found in almost any operations research book ([1], p. 573). However, the model in this chapter is considerably different from that in [1]. Items arrive for processing at the first machine M1 and then proceed through the next four machines M2., , , M5 until they reach the "finished" box. All queues in front of each machine are assumed to have unlimited capacity. Queue capacity will not be of importance in this problem so we did not show any queue in Fig. 14.1. Each machine can break down and this will certainly effect throughput, but all other unbroken machines continue to function. The probability that machine Mi breaks down is p_i, $1 \leq i \leq 5$. For each machine Mi the probability of a break down is modelled as follows: whenever an item arrives at the machine for processing, and goes into queue if the machine is busy, the machine can break down with probability p_i, $1 \leq i \leq 5$. If an item is being processed at break down it waits until the machine is fixed and then it continues its processing (no processing time is lost). There is a crew of k machine repair workers in the "repair" station in Fig. 14.1. The goal is to find the optimal k to minimize total costs.

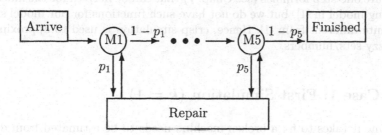

Fig. 14.1. Machine Servicing Problem

In Fig. 14.1 it looks like a broken machine is sent to the repair station for repair. This is not true. When a machine breaks down and there is an available repair worker, the repair worker is sent to the machine for repair.

James J. Buckley: *Simulating Fuzzy Systems*, StudFuzz **171**, 111–116 (2005)
www.springerlink.com   © Springer-Verlag Berlin Heidelberg 2005

The repair station is modelled as having k identical and parallel servers each server is a repair worker.

We may wish to add the possibility of defective items and limited queue capacity later but in this chapter we concentrate on finding k to minimize total costs.

First there is the hourly cost of a repair person, plus their needed tools and supplies, which we assume to be (including fringe benefits, etc.) 100 \$/hour. Next is the cost of lost production due to broken machines and down time. All simulations will be run until there are 50,000 units in the finished box. We are using the "swamping" method discussed in Chap. 7. Time units are in minutes. CLOCK gives the total number of minutes in the simulation. Let $W = (CLOCK)/(60)(8) = $ the number of 8 hour shifts in the simulation. If $D = $ is the total number of broken machines per simulation, if $T = $ the mean repair time (both being repaired and waiting to be repaired) per broken machine, and if $E = $ the cost (\$/hour) of machine down time, then total cost (TC) per 8 hour shift is

$$TC(k) = k(100)(8) + [D/W]TE . \tag{14.1}$$

In this equation D, W and T will all depend on k. We know that some of these numbers will be fuzzy. In the simulations for fixed k we will vary certain parameters through their α-cuts producing variability in D, W, and T. So these will all be fuzzy sets/numbers. Also, E is a very difficult number to estimate and we employ expert opinion (see Sects. 3.2, 12.1) to get the fuzzy estimator $\overline{E}$ for E. Therefore, out goal is to minimize

$$\overline{TC}(k) = 800k + \overline{[D/W]}\,\overline{T}\,\overline{E} , \tag{14.2}$$

for $k = 1, \ldots, 5$.

Now we simulate the fuzzy system to estimate some alpha-cuts of $\overline{TC}(k)$. There are one-step formulas (see Chap. 7) that can be derived for the machine servicing model in [1], but we do not have such functions for our model since it is quite different from [1]. Hence, crisp simulation is used to approximate the fuzzy sets/numbers.

14.2 Case 1: First Simulation ($k = 1$)

The time it takes to fix a broken machine needs to be estimated from data. Therefore (Chap. 3) we obtain a fuzzy estimator for the mean repair time. We will use the fuzzy exponential to model arrivals, the fuzzy uniform for machine (M1, ..., M5) processing times and the fuzzy normal for repair times. The fuzzy parameters values are all given in Table 14.1. However, we assume that the probability of a break down is known (crisp). We will use $\overline{E} = (125/150/175)$.

Table 14.1. Fuzzy/Crisp Probability Distributions for Fig. 14.1

Item	Distribution	Fuzzy/Crisp Parameters
Arrivals	Exponential	$\overline{\lambda} = (0.05/0.0833/0.1167)$
M1	Uniform	$\overline{a}_1 = (5/6/7), \overline{b}_1 = (9/10/11)$
M2	Uniform	$\overline{a}_2 = (5/6/7), \overline{b}_2 = (9/10/11)$
M3	Uniform	$\overline{a}_3 = (5/6/7), \overline{b}_3 = (9/10/11)$
M4	Uniform	$\overline{a}_4 = (5/6/7), \overline{b}_4 = (9/10/11)$
M5	Uniform	$\overline{a}_5 = (5/6/7), \overline{b}_5 = (9/10/11)$
Repair	Normal	$\overline{\mu}_r = (10/20/30), \sigma_r = 2.0$
p_1	Bernoulli	$p_1 = 0.10$
p_2	Bernoulli	$p_2 = 0.05$
p_3	Bernoulli	$p_3 = 0.07$
p_4	Bernoulli	$p_4 = 0.04$
p_5	Bernoulli	$p_5 = 0.08$

This case is for only one repair person. Note that the numbers for $\overline{\lambda}$ are given as rates (number of arrivals per time unit) so the reciprocal is the average time between arrivals. The values for $\overline{\mu}$ are the mean service times. From Chap. 8 we know how to choose λ and μ to approximate the end points of the α-cuts of $\overline{R}$ and $\overline{X}$. However, in this chapter we wish to do this for $\overline{[D/W]}$ and $\overline{T}$. We experimented with the λ, μ_r, a_i and b_i values to approximate the end points of the alpha-cuts. We ran simulations using max/min values of λ, μ_r, a_i and the b_i to see what combinations produce the max/min of $[D/W]$ and T. These results are in Tables 14.2 and 14.3. In the tables: (1) "min" means use the left end point of the interval for the α-cut of the fuzzy estimator; and (2) "max" means use the right end point. The information in Tables 14.2 and 14.3 can now be used to estimate alpha-cuts of $\overline{[D/W]}$ and $\overline{T}$ with the simulation results in Table 14.4. The "*" in Table 4.4 means that the simulation stopped before it reached 50,000 units

Table 14.2. Values of the Parameters for Min $[D/W]$, T

Min	λ	μ_r	a_i	b_i
$[D/W]$	min	min	min	min
T	min	min	min	min

Table 14.3. Values of the Parameters for Max $[D/W]$, T

Max	λ	μ_r	a_i	b_i
$[D/W]$	max	max	max	max
T	max	max	max	max

Table 14.4. Case 1 Simulation Alpha-Cuts of $\overline{[D/W]}, \overline{T}, \overline{TC}$ (1)

Item	$\alpha = 0$ Cut	$\alpha = 0.5$ Cut	$\alpha = 1$ Cut
$\overline{[D/W]}$	$[10.2, *]$	$[10.9, 12.6]$	11.7
$\overline{T}$	$[11.2, *]$	$[18.0, 40.4]$	27.0
$\overline{TC}(1)$	$[15,023.90, *]$	$[27,846.52, 83,937.7]$	48,088.01

Table 14.5. Case 2 Simulation Alpha-Cuts of $\overline{[D/W]}, \overline{T}, \overline{TC}$ (2)

Item	$\alpha = 0$ Cut	$\alpha = 0.5$ Cut	$\alpha = 1$ Cut
$\overline{[D/W]}$	$[10.2, 13.4]$	$[10.8, 12.5]$	11.9
$\overline{T}$	$[10.0, 32.2]$	$[15.1, 26.0]$	20.5
$\overline{TC}(2)$	$[12,827.38, 76,949.55]$	$[24,068.04, 52,797.83]$	38,149.66

Table 14.6. Case 3 Simulation Alpha-Cuts of $\overline{[D/W]}, \overline{T}, \overline{TC}$ (3)

Item	$\alpha = 0$ Cut	$\alpha = 0.5$ Cut	$\alpha = 1$ Cut
$\overline{[D/W]}$	$[10.1, 13.6]$	$[10.9, 12.6]$	11.7
$\overline{T}$	$[10.0, 30.2]$	$[15.0, 25.0]$	20.0
$\overline{TC}(3)$	$[15,000.00, 74,405.54]$	$[24,978.79, 53,903.68]$	37,604.08

Table 14.7. Case 4 Simulation Alpha-Cuts of $\overline{[D/W]}, \overline{T}, \overline{TC}$ (4)

Item	$\alpha = 0$ Cut	$\alpha = 0.5$ Cut	$\alpha = 1$ Cut
$\overline{[D/W]}$	$[10.1, 13.6]$	$[10.8, 12.6]$	11.7
$\overline{T}$	$[10.0, 30.0]$	$[15.0, 25.0]$	20.0
$\overline{TC}(4)$	$[15,793.70, 74,544.91]$	$[25,470.36, 54,527.85]$	38,300.14

in the finished box because it ran out of memory (Chap. 6). We calculated the α-cuts of $\overline{TC}(k)$ using more decimal places in the alpha-cuts of $\overline{[D/W]}$ and $\overline{T}$ than the rounded off values in Tables 14.4–14.8.

The value of CLOCK and the run time varies with the values of the parameters. However, the simulation was usually around 1400 eight hour shifts and the run time varied from 1 to 2 seconds.

We show the graphs of $\overline{TC}(k)$, $k = 1, \ldots, 5$, in Fig. 14.2.

14.3 Case 2: Second Simulation ($k = 2$)

We change $k = 1$ to $k = 2$ and simulate the fuzzy system again with results in Table 14.5.

Table 14.8. Case 5 Simulation Alpha-Cuts of $\overline{[D/W]}$, $\overline{T}$, $\overline{TC}$ (5)

Item	$\alpha = 0$ Cut	$\alpha = 0.5$ Cut	$\alpha = 1$ Cut
$\overline{[D/W]}$	[10.1, 13.6]	[10.8, 12.6]	11.6
$\overline{T}$	[10.0, 30.0]	[15.0, 25.0]	20.0
$\overline{TC}(5)$	[16,639.71, 75,376.04]	[26,270.36, 55,086.05]	38,768.68

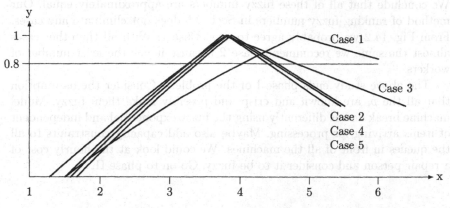

Fig. 14.2. $\overline{TC}(k)$ (Dollars) in Cases 1–5. Scale = $x(10,000)$

14.4 Case 3: Third Simulation ($k = 3$)

Now there are three repair workers. Simulation produces the α-cuts in Table 14.6.

14.5 Case 4: Fourth Simulation ($k = 4$)

The fourth simulation is for four repair workers and the alpha-cuts are shown in Table 14.7.

14.6 Case 5: Fifth Simulation ($k = 5$)

The last simulation is when there is five repair persons with α-cuts in Table 14.8.

14.7 Summary

We see that $\overline{[D/W]}$ is fairly constant for $k = 1, \ldots, 5$ because the p_i, the probability of machine Mi breaking down, was assumed known and it was not fuzzy. Also, $\overline{T}$, down time for a broken machine, was fairly constant for $k = 2, 3, 4, 5$. The result is that the term $\overline{[D/W]} \, \overline{T} \, \overline{E}$ in $\overline{TC}(k)$ was approximately level (constant) for $k = 2, 3, 4, 5$ and shows its largest change in going from $k = 1$ to $k = 2$ since $\overline{T}$ decreased. Now add the term $800k$ to get all of $\overline{TC}(k)$ with their graphs in Fig. 14.2.

In Fig. 14.2 all the fuzzy numbers intersect above the horizontal line at 0.8, see Fig. 2.4. The base of these fuzzy numbers is like a 99% confidence interval. We conclude that all of these fuzzy numbers are approximately equal. Our method of ranking fuzzy numbers in Sect. 2.5 does not eliminate any cases. From Fig. 14.2 we probably agree to drop Case 1. With all the other cases almost the same we recommend Case 2 because it has the least number of workers.

The above study ends phase I of the problem. Consider the assumption that all the p_i are known and crisp and possibly make them fuzzy. Model machine break downs differently using the fuzzy exponential and independent of items arriving for processing. Maybe also add capacity constraints to all the queues in front of all the machines. We could look at the hourly cost of a repair person and consider it to be fuzzy. Go on to phase II.

Reference

1. H.A. Taha: Operations Research, Fifth Edition, Macmillan, N.Y., 1992.

15 Life Insurance: New Account Model

15.1 Introduction

This problem was derived from an example in [1]. The system is shown in
Fig. 15.1. This figure shows how this insurance company handles an application for an insurance policy. New account applications arrive at the rate of λ
per unit time. We use the exponential to model time between arrivals with
$1/\lambda$ the mean time between arrivals. A new account first goes to the credit
check (CC in Fig. 15.1) station. All service times will be modelled by the
normal distribution. The clerk at CC can reject the application, with probability p_1, or not reject and send it on to "SPLIT". The SPLIT operation
creates a "clone" (an identical copy of the application) and sends one copy to
RR (risk rating) and the other to M (medical check). The SPLIT operation
is assumed to take no time, or there is no delay at SPLIT. The medical check
station is staffed by medical doctors who review the medical history of the
applicant and the risk rating station has clerks who evaluate the company's
risk in the policy. After M and RR the two copies come together at JOIN
and proceed as one file.

The two copies of the application can arrive at JOIN at different times.
The first copy to arrive at JOIN must wait there until its clone arrives for both
to be joined into one file. We assume that there is no significant time delay in
the joining operation once both copies are present. Once joined the file goes
to the review station (R in Fig. 15.1). A clerk at R may reject the application
with probability p_2, or send it on to the pricing station (P in the figure). After
pricing the application proceeds to the paper work station (PW) where all
the relevant information is assembled into an insurance policy and the new
account ends up in the approved box. It is assumed that all queues have
unlimited capacity and have not been shown in Fig. 15.1.

Management wants the time it takes to process an application to be minimized. They would like to investigate the staffing of the various work stations
to reduce R = mean time in the system for an any (rejected/approved) application. However, as they add more clerks to CC, RR, R and P the total
costs increase. They would like to keep the costs within some target budget
while minimizing R.

Since we do not have any one-step formulas (Chap. 7) for R in this model
we will simulate it to estimate α-cuts of $\overline{R}$. The costs associated with this

James J. Buckley: *Simulating Fuzzy Systems*, StudFuzz **171**, 117–121 (2005)
www.springerlink.com   © Springer-Verlag Berlin Heidelberg 2005

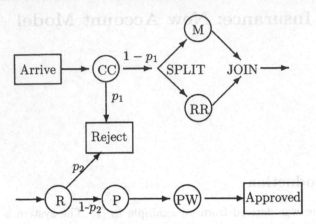

Fig. 15.1. New Account Model

system will be the personnel costs for one eight hour shift. This cost will be crisp and not fuzzy. Personnel costs are made up from the secretaries, clerks and doctors. These hourly costs, including fringe benefit expenses, etc. are in Table 15.1. The budget constraint is $B per eight hour shift. So we wish to $min\overline{R}$ subject to $Costs \leq B$. All simulations will be run until 50,000 new applications have been processed. Simulation time is in minutes. We need to run the simulation a long time so as not to bias R with the start-up results of all service stations empty (the swamping method Chap. 7). A simulation run of 50,000 is equal to approximately 300 eight hour shifts for the new account department. Also, each simulation runs takes around 2 seconds.

15.2 Case 1: First Simulation

This case is to see how the system will perform with minimal staffing. Minimal staffing is to use only one clerk/secretary/doctor at their service station. From the data collected on the new accounts department we estimate the arrival rates and the service times. These will be fuzzy estimators. We discussed

Table 15.1. Staffing Costs in the New Account Department

Station	Personnel	Cost($/hour)
CC	Clerk	65
M	Doctor	150
RR	Clerk	65
R	Clerk	65
P	Clerk	65
PW	Secretary	50

Table 15.2. Fuzzy Probability Distributions for Fig. 15.1

Item	Distribution	Fuzzy Parameters
Arrivals	Exponential	$\overline{\lambda} = (0.0286/0.0333/0.0400)$
CC	Normal	$\overline{\mu}_{cc} = (15/20/25), \sigma_{cc} = 3.0$
M	Normal	$\overline{\mu}_m = (25/30/35), \sigma_m = 5.0$
RR	Normal	$\overline{\mu}_{rr} = (10/15/20), \sigma_{rr} = 2.0$
R	Normal	$\overline{\mu}_r = (15/20/25), \sigma_r = 3.0$
P	Normal	$\overline{\mu}_p = (5/10/15), \sigma_p = 1.0$
PW	Normal	$\overline{\mu}_{pw} = (25/30/35), \sigma_{pw} = 5.0$
p_1	Bernoulli	$p_1 = (0.05/0.10/0.15)$
p_2	Bernoulli	$p_2 = (0.03/0.05/0.07)$

in Chap. 8 that we will use only a crisp point estimator for the standard deviation of the fuzzy normal distribution. The fuzzy probability distributions for the system are defined in Table 15.2. The percent of rejects (p_1, p_2) varies because of the different polices requested and the variability in the applications. So, the p_i will also be fuzzy. Note that the numbers for $\overline{\lambda}$ are given as rates (number of arrivals per time unit) so the reciprocal is the average time between arrivals. The values for $\overline{\mu}$ are the mean service times. Using Chap. 8 we know how to choose the λ and the μ to approximate the end points of the α-cuts of $\overline{R}$. In general we use: (1) $min\lambda$ and $min\mu$ for $minR$; but (2) $max\lambda$ and $max\mu$ for $maxR$. We need to experiment with the p_i values to see which ones to use for $max/minR$. The results are in Tables 15.3 and 15.4. We will use $B = \$8,280$ per eight hour shift. Notice that the minimum mean time in the system for a new account that is approved is 110 minutes, which is calculated using only one worker per station and the center (membership value one) of the fuzzy numbers for service time through the route CC-M-R-P-PW.

The simulation results are in Table 15.5. The * for the right end point of $\overline{R}[0]$ in Table 15.5 means we ran out memory even for a run of size 1000. Also, the superscript of * for the right end point of $\overline{R}[0.5]$ means a run size of

Table 15.3. Values of the Parameters for Min R

Min	λ	μ	p_1	p_2
R	min	min	max	max

Table 15.4. Values of the Parameters for Max R

Max	λ	μ	p_1	p_2
R	max	max	min	min

Table 15.5. Case 1 Simulation Alpha-Cuts of $\overline{R}$

Item	$\alpha = 0$ Cut	$\alpha = 0.5$ Cut	$\alpha = 1$ Cut
$\overline{R}$	$[94.6, *]$	$[128.6, 1003.5^*]$	263.8

1000 since larger values ran out of memory. With minimal staffing the costs per eight hour shift is $3,680 well under budget.

We show the graphs of $\overline{R}$ in Fig. 15.2.

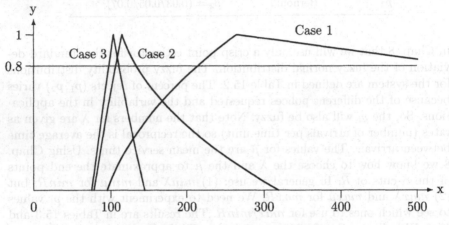

Fig. 15.2. $\overline{R}$ (Minutes) in Cases 1–3

15.3 Case 2: Second Simulation

In Case 1, from the standard simulation output file, we see that the utilization of M is 0.903, of PW is 0.858 and of CC is 0.669. So, let us try two doctors in M, two clerks in CC and two secretaries in PW. The new total cost is $5800 per eight hour shift, still under budget. Using this staffing, the simulation results are in Table 15.6. Let $\overline{R}_i$ be the fuzzy value of R in Case $i = 1, 2, 3$. We get that $\overline{R}_2 < \overline{R}_1$, see Sect. 2.5, from Fig. 15.1.

We do not expect to make further substantial reductions in $\overline{R}$ since $\overline{R}[1] = 113.9$ and (as discussed above) the minimum $\alpha = 1$ mean value of R is 110. However, the 110 figure was for one person at each service station and we added more workers. Since we have some money left in the budget let us try to reduce $\overline{R}$ further.

Table 15.6. Case 2 Simulation Alpha-Cuts of $\overline{R}$

Item	$\alpha = 0$ Cut	$\alpha = 0.5$ Cut	$\alpha = 1$ Cut
$\overline{R}$	[77.0, 317.7]	[93.3, 146.6]	113.9

15.4 Case 3: Third Simulation

We have found a way to spend all the money in the budget. It is: (1) three clerks in CC; (2) two doctors in M; (3) two clerks in RR, R, P; and (4) three secretaries in PW. The total cost is \$8280 per eight hour shift and all the money in the budget is gone. We are curious to see what happens with this staffing. Simulation results are in Table 15.7 and the graph of $\overline{R}_3$ (Case 3) is in Fig. 15.2.

Table 15.7. Case 3 Simulation Alpha-Cuts of $\overline{R}$

Item	$\alpha = 0$ Cut	$\alpha = 0.5$ Cut	$\alpha = 1$ Cut
$\overline{R}$	[74.0, 143.3]	[87.7, 120.2]	102.9

15.5 Summary

This chapter was about staffing work stations in a new account department in an insurance company. The objective was to minimize $\overline{R}$, the fuzzy number for the time an application (rejected/approved) spends in the system, subject to a fixed budget constraint. Three cases were studied with their results shown in Fig. 15.2. The base of these fuzzy numbers is like a 99% confidence interval. We see that $\overline{R}_3 \approx \overline{R}_2 < \overline{R}_1$. Having only these three cases we first recommend Case 2 because it spends less money than Case 3. However, there is some worry about the large value of the right end point of $\overline{R}_2[0]$ which implies that we go with Case 3. Which case would you choose?

Reference

1. Getting Started SIMPROCESS, CACI Products Company, San Diego, CA, 2004.

16 Inventory Control I

16.1 Introduction

This single item multi-period inventory model is shown in Fig. 16.1. This problem will be expanded and also studied in the next chapter. These inventory control models may be found in most operation research books ([2], Chap. 14) and is chapter is modelled after the paper [1].

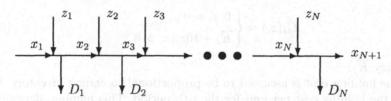

Fig. 16.1. Inventory Control Problem I

We first describe the inventory problem in Fig. 16.1. The incoming inventory x_1 will always be a given real number. The variables are the $z_1, \ldots, z_N$ which are the amounts we are to order each period. The z_i, $1 \leq i \leq N$, are non-negative integers. The D_i represent the demand in the i-th period and the x_i, $2 \leq i \leq N$, stands for the outgoing inventory, which will be the starting inventory for the next period, in the i-th period. So we must have

$$x_{i+1} = x_i + z_i - D_i, \ 1 \leq i \leq N, \tag{16.1}$$

as long as $x_{i+1} \geq 0$. If $x_i + z_i - D_i < 0$, then we set $x_{i+1} = 0$.

There are only N periods and at the end of the planning horizon management would like final inventory to be close to zero. That is, a goal is to obtain $x_{N+1} = 0$. We will model demand using the fuzzy normal distribution so let the fuzzy estimator of the mean be $\overline{\mu}$ with crisp standard deviation σ. Since $\overline{D}_i$ will be fuzzy, then the $\overline{x}_{i+1}$, $i \geq 1$, in (16.1) will also be fuzzy, and hence $\overline{x}_{N+1}$ is also fuzzy. A goal is to $min\overline{x}_{N+1}$.

There are two basic assumptions in this chapter which will be relaxed in the next chapter. First, we assume that there is zero delivery lag with instant replenishment at the start of each period. During the i-th period we place

James J. Buckley: *Simulating Fuzzy Systems*, StudFuzz **171**, 123–129 (2005)
www.springerlink.com   © Springer-Verlag Berlin Heidelberg 2005

an order for z_i units which all arrive at the start of the next period. Also, there is no backlogging. What this means is that if $x_i + z_i - D_i < 0$, demand exceeds supply, these orders are lost and we do not carry these orders over to the next period's demand. Another assumption, used in both chapters, is that future money is treated as equal to present money and we will not use the present value (discount future money back to the present).

The second goal is to minimize total inventory cost over the N periods. This cost is made up of four components: (1) purchase cost; (2) ordering cost; (3) holding cost; and (4) shortage cost. We now discuss these in detail.

We assume that we are to buy the item and we do not produce it ourselves. There may, or may not, be price breaks. For simplicity let us assume that there are no price breaks and it costs \$10 per unit to purchase one unit. The ordering cost K_i in the i-th period is the cost of placing the order, checking up on the order and putting the items into inventory when they arrive. This number is always difficult to estimate so we will model it using a fuzzy number (obtained from expert opinion, Sects. 3.2, 12.1). Then the total cost of obtaining z_i units at the start of the i-th period is

$$\overline{C}_i(z_i) = \begin{cases} 0, z_i = 0 \,, \\ \overline{K}_i + 10z_i, z_i > 0 \,, \end{cases} \tag{16.2}$$

for fuzzy $\overline{K}_i$.

The holding cost is assumed to be proportional to ending inventory. Let h_i be the holding cost per unit for the i-th period. This number, depending on interest on invested capital, depreciation, etc., is very difficult to estimate exactly, so it will also be fuzzy. Through expert opinion let out fuzzy estimator of the holding cost be $\overline{h}_i$. Hence the holding cost for the i-th period is

$$\overline{H}_i(z_i) = \overline{h}_i(\overline{x}_i + z_i - D_i) \,, \tag{16.3}$$

as long as ending inventory is positive, otherwise the holding cost is zero.

When we run out of inventory and demand exceeds supply, we have lost sales and there is a penalty. Let p_i be the penalty cost per lost sale which includes the money lost in the sale plus some amount for the loss of customer "good will". This is another value almost impossible to estimate precisely, so we will have a fuzzy estimator $\overline{p}_i$ for the lost sales penalty. The lost sales cost for the i-th period is

$$\overline{L}_i(z_i) = \overline{p}_i(\overline{D}_i - \overline{x}_i - z_i) \,, \tag{16.4}$$

as long as lost sales is positive, otherwise it is zero.

The total inventory cost for the N periods is

$$\overline{TC} = \sum_{i=1}^{N} [\overline{C}_i(z_i) + \overline{H}_i(z_i) + \overline{L}_i(z_i)] \,. \tag{16.5}$$

The goal is to find the z_i to $min\overline{TC}$.

Management agrees to consider the optimization problem

$$min\overline{Z} = min\left\{\tau_1[\overline{TC}/1000] + \tau_2\overline{x}_{N+1}\right\}, \tag{16.6}$$

for selected $\tau_i > 0$, $\tau_1 + \tau_1 = 1$. Equation 16.6 combines the two fuzzy numbers for the two goals into one fuzzy number and then we rank the resulting $\overline{Z}$, obtained from choices for the z_i, using Sect. 2.5. $\overline{TC}$ will be from around 5,000 to 10,000 while x_{N+1} is in the range 0 to 50. We scaled TC by dividing by 1000 so they both get approximately in the same range.

When this inventory control problem is crisp it is usually solved using dynamic programming. The fuzzy case was solved using an evolutionary algorithm in [1]. Here, we will use simulation.

16.2 Case 1: First Simulation

Let us assume that there are five ($N = 5$) periods. All simulations will be for 50,000 runs through the five periods. We are using the "swamping" method discussed in Chap. 7, where we run the simulation sufficiently long so that the results will not be biased by the start up conditions. For a given α-cut, after the 50,000 runs, we may find the average values of TC and x_6. For example, the program calculates the total inventory cost for each run through the five periods, and then adds these up over all 50,000 runs to get TOTAL. Then $TOTAL/50,000$ is the mean presented in the following tables. We estimate the end points of $\overline{x}_6[\alpha]$ the same way. All run times were around 3 seconds. For simplicity we will use the same values of $\overline{\mu}$, σ, $\overline{K}_i$, $\overline{h}_i$ and $\overline{p}_i$ all periods and their values are in Table 16.1.

Table 16.1. Fuzzy Numbers in Inventory Control I

Item	Fuzzy/Crisp Parameter	Value
Demand	$\overline{\mu}, \sigma$	$\overline{\mu} = (70/76/82), \sigma = 10$
Ordering Cost	$\overline{K}$	$(92/98/104)$
Holding Cost	$\overline{h}$	$(15/20/25)$
Shortage Cost	$\overline{p}$	$(25/30/35)$

Case 1 is where the orders are $z_1 = 61$, $z_2 = z_3 = z_4 = z_5 = 76$. Here we order equal to the central value of $\overline{\mu}$ and $z_1 = 61$ since $x_1 = 15$. In this chapter we use $x_1 = 15$. The total cost of buying all these units (the z_i values) is \$4140 using $K = \$98$, so the absolute minimum $\alpha = 1$ total inventory cost in Case 1 is \$4140.

Next we need to know how to choose the values of μ, K, h and p for the simulation to approximate the end points of the intervals for the alpha-cuts of $\overline{TC}$ and $\overline{x}_6$. We discovered that this will depend on the values we use for the

Table 16.2. Cases 1, 3 Values of the Parameters for Min TC, x_{N+1}

Min	μ	K	h	p
TC	max	min	min	min
x_{N+1}	max	max/min	max/min	max/min

Table 16.3. Cases 1, 3 Values of the Parameters for Max TC, x_{N+1}

Max	μ	K	h	p
TC	min	max	max	max
x_{N+1}	min	max/min	max/min	max/min

Table 16.4. Case 1 Simulation Alpha-Cuts of $\overline{TC}$, $\overline{x}_{N+1}$

Item	$\alpha = 0$ Cut	$\alpha = 0.5$ Cut	$\alpha = 1$ Cut
$\overline{TC}$	$[5148, 6887]$	$[5169, 5966]$	5407
$\overline{x}_{N+1}$	$[3, 33]$	$[7, 22]$	13

Table 16.5. Case 2 Values of the Parameters for Min TC, x_{N+1}

Min	μ	K	h	p
TC	min	min	min	min
x_{N+1}	max	max/min	max/min	max/min

z_i. The results are shown in Tables 16.2–16.3 and 16.5–16.6. The max/min values of x_6 do not depend on K, h or p but only on μ. The max/min values of TC depend on μ, K, h and p and the results can vary with the z_i values.

This problem is different from those in the previous chapters. The inventory control problem is not a queuing system. We need not worry about simulating long enough to be in steady-state. However, we do assume that there are continuous one-step functions, see Chap. 7, so that $TC = F(\mu, K, h, p, \ldots)$ and $x_{N+1} = G(\mu, x_1, z_1, \ldots)$. Then by the extension principle $\overline{TC} = F(\overline{\mu}, \overline{K}, \overline{h}, \overline{p}, \ldots)$ and $\overline{x}_{N+1} = G(\overline{\mu}, x_1, z_1, \ldots)$. The crisp simulation is to approximate the alpha-cuts of these fuzzy functions since we

Table 16.6. Case 2 Values of the Parameters for Max TC, x_{N+1}

Max	μ	K	h	p
TC	max	max	max	max
x_{N+1}	min	max/min	max/min	max/min

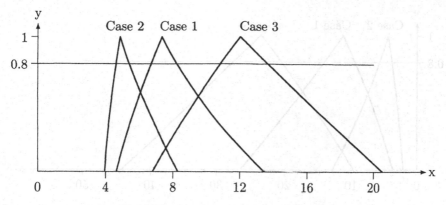

Fig. 16.2. $\overline{Z}$, $\tau_1 = 3/4$, $\tau_2 = 1/4$, in Cases 1, 2, 3

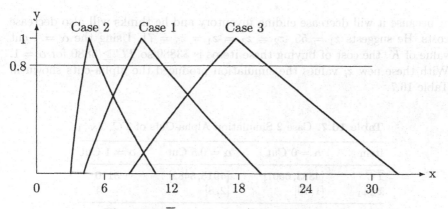

Fig. 16.3. $\overline{Z}$, $\tau_1 = \tau_2 = 1/2$, in Cases 1, 2, 3

do not know exact formulas for these functions. The simulation results are in Table 16.4.

Management agrees to consider three cases in computing $\overline{Z}$: (1) weight costs three times as much as ending inventory; (2) weight both equally; and (3) weight ending inventory three times as much as costs. We show the graphs of $\overline{Z}$ for these cases in Figs. 16.2–16.4.

16.3 Case 2: Second Simulation

There are two engineers who work for this company and they present their completely different plans on how to solve this inventory control problem. The first engineer says that since lost sales penalty (p) is much larger than holding cost (h), we should increase the z_i to decrease lost sales. This will be Case 3 considered below. The other engineer says that we should decrease the

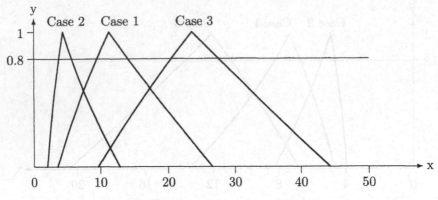

Fig. 16.4. $\overline{Z}$, $\tau_1 = 1/4$, $\tau_2 = 3/4$, in Cases 1, 2, 3

z_i because it will decrease ending inventory and he thinks will also decrease costs. He suggests $z_1 = 55$, $z_2 = z_3 = z_4 = z_5 = 71$. Using the $\alpha = 1$ cut value of $\overline{K}$, the cost of buying these items is \$3880 so $TC \geq 3880$ for $\alpha = 1$. With these new z_i values the simulation produced the alpha-cuts shown in Table 16.7.

Table 16.7. Case 2 Simulation Alpha-Cuts of $\overline{TC}$, $\overline{x}_{N+1}$

Item	$\alpha = 0$ Cut	$\alpha = 0.5$ Cut	$\alpha = 1$ Cut
$\overline{TC}$	[4883, 6007]	[4915, 5482]	5110
$\overline{x}_{N+1}$	[1, 15]	[2, 8]	4

16.4 Case 3: Third Simulation

This engineer wants to use $z_1 = 66$, $z_2 = z_3 = z_4 = z_5 = 81$. The cost of buying the items is \$4390 for $K = 98. Hence, $TC \geq 4390$ when alpha is one. Using these values for the z_i the simulation results are in Table 16.8.

Table 16.8. Case 3 Simulation Alpha-Cuts of $\overline{TC}$, $\overline{x}_{N+1}$

Item	$\alpha = 0$ Cut	$\alpha = 0.5$ Cut	$\alpha = 1$ Cut
$\overline{TC}$	[5302, 8695]	[5675, 7376]	6362
$\overline{x}_{N+1}$	[11, 56]	[19, 42]	29

16.5 Summary

Let $\overline{Z}_{ij}$, from (16.6), be the value of $\overline{Z}$ in Case $i = 1, 2, 3$ and: (1) $j = 1$ for $\tau_1 = 3/4$, $\tau_2 = 1/4$; (2) $j = 2$ if $\tau_1 = \tau_2 = 1/2$; and (3) $j = 3$ when $\tau = 1/4$, $\tau_2 = 3/4$. The base of these fuzzy numbers is like a 99% confidence interval. We see from Figs. 16.2–16.4 that $\overline{Z}_{2j} < \overline{Z}_{1j} < \overline{Z}_{3j}$ for $j = 1, 2, 3$. Of the three cases considered, Case 2 is the best. Management will look at more cases but let us go on to the next chapter and add delivery lag and backlogging to the inventory model. It might be fun to guess other values for the z_i to see if we can do better than Case 2. What is the "optimal" solution?

References

1. J.J. Buckley, T. Feuring and Y. Hayashi: Solving Fuzzy Problems in Operations Research: Inventory Control, Soft Computing, 7(2002) 121-129.
2. H.A. Taha: Operations Research, Fifth Edition, Macmillan, N.Y., 1992.

17 Inventory Control II

17.1 Introduction

This chapter continues the inventory control problem studied in the previous chapter. The new system is shown in Fig. 17.1. We have added two things to the model in Chap. 16: (1) backlogging (backordering) which is shown in Fig. 17.1 as an increase in demand to $D_i + s_i$, $2 \leq i \leq N$; and (2) delivery lag shown as a 80% chance that an order arrives at the start of a period and a 20% chance that it arrives at the start of the next period. We assume that these probabilities, the 0.80 and 0.20, are known and will not be fuzzy. A fuzzy inventory problem similar to this has been considered in [1].

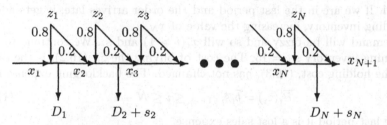

Fig. 17.1. Inventory Control Problem II

Now we discuss the changes in the inventory control model of Chap. 16. The basic equation for the system is

$$x_{i+1} = x_i + z_i - D_i, \ 1 \leq i \leq N, \tag{17.1}$$

as long as $x_{i+1} \geq 0$. If $x_i + z_i - D_i < 0$, then set $x_{i+1} = 0$ and define $s_{i+1} = D_i - (x_i + z_i)$. When demand exceeds supply, the backorder amount s_{i+1} is added to next periods demand. If $D_i > x_i + z_i$ the excess demand is not lost, the customer is told to come back next period, which we assume they all will do, where we will try to fill their order. So, if $s_{i+1} > 0$, the demand next period is $D_{i+1} + s_{i+1}$ as shown in Fig. 17.1. If demand does not exceed supply, s_{i+1} is set to zero and next periods demand remains D_{i+1}. So, the demand for the i-th period is D_i if $s_i = 0$ and $D_i + s_i$ if $s_i > 0$. Backordering

James J. Buckley: *Simulating Fuzzy Systems*, StudFuzz **171**, 131–136 (2005)
www.springerlink.com   © Springer-Verlag Berlin Heidelberg 2005

has an extra cost of $b\$/unit$. This number b is generally hard to estimate so we will model it as a fuzzy number given in Table 17.1. Table 17.1 has all the fuzzy numbers to be used in this chapter and they are all the same as in Chap. 16 with the addition of b. In the simulations we will use the same values of $\overline{\mu}$, $\overline{K}$, $\overline{h}$, and $\overline{b}$ in all periods. We perform backodering, as needed, until the last period when s_{N+1}, assuming it is positive, becomes lost sales with penalty $\overline{p}$.

Table 17.1. Fuzzy Numbers in Inventory Control II

Item	Fuzzy Parameter	Value
Demand	$\overline{\mu}, \sigma$	$\overline{\mu} = (70/76/82), \sigma = 10$
Ordering Cost	$\overline{K}$	$(92/98/104)$
Holding Cost	$\overline{h}$	$(15/20/25)$
Backlogging Cost	$\overline{b}$	$(20/25/30)$
Shortage Cost	$\overline{p}$	$(25/30/35)$

The possibility of a delivery lag is shown in Fig. 17.1. An order is placed for z_i units and we have probability 0.8 that it will arrive at the start of the i-th period but a probability of 0.2 that it arrives at the start of the next period. If we are in the last period and the order arrives late, it gets added to ending inventory, increasing the value of x_{N+1}.

Demand will be fuzzy and so will $x_i(i > 1)$ and s_i. We continue to use beginning inventory $x_1 = 15$. The cost of buying the z_i, (16.2), is the same and the holding cost, (16.3), has not changed. The backlogging expense is

$$\overline{B}_i(z_i) = \overline{b}_i \overline{s}_{i+1}, \ 1 \leq i \leq N - 1 .$$ (17.2)

In the last period it is a lost sales expense

$$\overline{L}_N(z_N) = \overline{p} \ \overline{s}_{N+1} .$$ (17.3)

Recall that $s_{i+1} = D_i - (x_i + z_i)$ if it is positive, otherwise it is zero.

The total inventory cost over the N periods is

$$\overline{TC} = \sum_{i=1}^{N} [\overline{C}_i(z_i) + \overline{H}_i(z_i) + \overline{B}_i^*(z_i)] ,$$ (17.4)

where $\overline{B}^*$ means to switch to $\overline{L}_N$ for the last period.

The goals are the same of $min\overline{x}_{N+1}$ and $min\overline{TC}$ together with keeping the final lost sales (s_{N+1}) a minimum. The final lost sales is the amount, if any, demand exceeds supply in the final period. The three fuzzy goals are combined into one fuzzy goal as in (16.6) which produces

$$min\overline{Z} = min\left\{\tau_1[\overline{TC}/1000] + \tau_2\overline{x}_{N+1} + \tau_3\overline{s}_{N+1}\right\} ,$$ (17.5)

for $\tau_i > 0$ and $\tau_1 + \tau_2 + \tau_3 = 1$.

17.2 Case 1: First Simulation

We will use five periods ($N = 5$) as in Chap. 16. All simulations will be for
50,000 runs through the five periods. We calculate the mean values for TC,
x_6 and s_6 as discussed in Sect. 16.2. The run times now turned out to be
between 3 and 4 seconds.

We start with using the order sizes equal to those in Case 2 of Chap. 16.
These values for the z_i gave the best results in that chapter. They are $z_1 = 55$,
$z_2 = z_3 = z_4 = z_5 = 71$. The $\alpha = 1$ minimum total inventory cost is \$3880.
We know how to pick the values of μ, K, h and p to get the max/min values
for TC and x_6 from Chap. 16. Now we have added b and s_6. Clearly b does
not effect x_6 and we should use $max\ b$ ($min\ b$) for $max\ TC$ ($min\ TC$).
Also, the values of K, h, b and p will not effect s_6. Experimenting with
max/min μ and s_6 we determine the results shown in Tables 17.2 and 17.3.
We know, from Chap. 16, that the values of the z_i may effect the values
of μ for max/min TC. Using these tables we can run the simulations to
approximate the needed α-cuts with results in Table 17.4.

Table 17.2. Cases 1–3 Values of the Parameters for Min TC, x_{N+1}, s_{N+1}

Min	μ	K	h	b	p
TC	min	min	min	min	min
x_{N+1}	max	max/min	max/min	max/min	max/min
s_{N+1}	min	max/min	max/min	max/min	max/min

Table 17.3. Cases 1–3 Values of the Parameters for Max TC, x_{N+1}, s_{N+1}

Max	μ	K	h	b	p
TC	max	max	max	max	max
x_{N+1}	min	max/min	max/min	max/min	max/min
s_{N+1}	max	max/min	max/min	max/min	max/min

Table 17.4. Case 1 Simulation Alpha-Cuts of $\overline{TC}$, $\overline{x}_{N+1}$, $\overline{s}_{N+1}$

Item	$\alpha = 0$ Cut	$\alpha = 0.5$ Cut	$\alpha = 1$ Cut
$\overline{TC}$	[6183, 11,441]	[6923, 9592]	8065
$\overline{x}_{N+1}$	[0, 9]	[0, 4]	1
$\overline{s}_{N+1}$	[19, 70]	[29, 55]	41

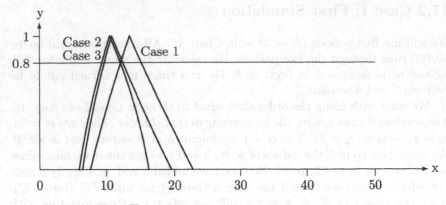

Fig. 17.2. $\overline{Z}$, $\tau_1 = 3/5$, $\tau_2 = 1/5$, $\tau_3 = 1/5$, in Cases 1, 2, 3

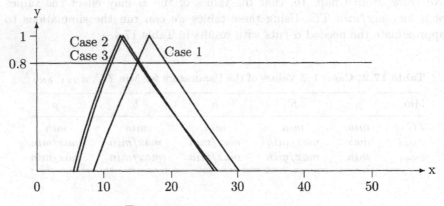

Fig. 17.3. $\overline{Z}$, $\tau_1 = 1/3$, $\tau_2 = 1/3$, $\tau_3 = 1/3$, in Cases 1, 2, 3

Graphs of the fuzzy goal function $\overline{Z}$, for selected values of the τ_i, are shown in Figs. 17.2–17.4.

17.3 Case 2: Second Simulation

The engineers feel that the value of $\overline{s}_6$ is too large. A large value of $\overline{s}_6$ inflates $\overline{TC}$ through (17.3). So they agree to increase the z_i values to $z_1 = 61$, $z_2 = 78$, $z_3 = 79$, $z_4 = 80$, $z_5 = 82$. The sum of the previous z_i values was 339 and these new z_i values add up to 380 an increase of 41. This value of 41 is the $\alpha = 1$ cut value of $\overline{s}_6$ in Table 17.4. The $\alpha = 1$ minimum inventory cost is \$4290 which is the cost of buying the items using $K = \$98$.

We know how to choose μ for the max/min x_6 and for max/min s_6 as shown in Tables 17.2 and 17.3. With these new values of the z_i we need to experiment with the μ values to obtain max/min TC. The final results are

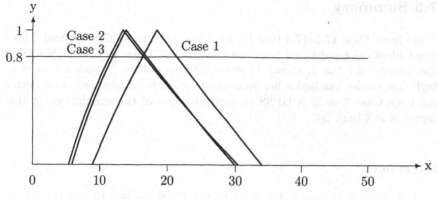

Fig. 17.4. $\overline{Z}$, $\tau_1 = 1/5$, $\tau_2 = 2/5$, $\tau_3 = 2/5$, in Cases 1, 2, 3

the same as those in Tables 17.2 and 17.3. Hence, we may proceed with the simulation with the alpha-cuts shown in Table 17.5. Now consider the graphs of $\overline{Z}$ for Case 2 in Figs. 17.2–17.4.

17.4 Case 3: Third Simulation

It now becomes difficult to find z_i values to significantly reduce a $\overline{Z}$ value. Considering the situation were inventory costs are weighted three times as high as final inventory (x_6) and final lost sales (s_6), or $\tau_1 = 3/5$, $\tau_2 = \tau_3 = 1/5$ in Fig. 17.2, the engineers come up with $z_1 = 61$, $z_2 = 78$, $z_3 = 77$, $z_4 = 75$, $z_5 = 66$. We will now try these values. Their sum is 357 and the cost of buying them ($K = \$98$) is 4060. For these new z_i values, the values of μ for $max/min\ TC$ are determined and the results are the same as in Tables 17.2–17.3. The simulation results are in Table 17.6.

Table 17.5. Case 2 Simulation Alpha-Cuts of $\overline{TC}$, $\overline{x}_{N+1}$, $\overline{s}_{N+1}$

Item	$\alpha = 0$ Cut	$\alpha = 0.5$ Cut	$\alpha = 1$ Cut
$\overline{TC}$	[6906, 9061]	[7071, 7978]	7365
$\overline{x}_{N+1}$	[3, 36]	[7, 25]	15
$\overline{s}_{N+1}$	[8, 34]	[11, 24]	16

Table 17.6. Case 3 Simulation Alpha-Cuts of $\overline{TC}$, $\overline{x}_{N+1}$, $\overline{s}_{N+1}$

Item	$\alpha = 0$ Cut	$\alpha = 0.5$ Cut	$\alpha = 1$ Cut
$\overline{TC}$	[6408, 9537]	[6674, 8151]	7207
$\overline{x}_{N+1}$	[0, 19]	[1, 10]	4
$\overline{s}_{N+1}$	[10, 52]	[16, 38]	26

17.5 Summary

We see from Figs. 17.2–17.4 that for all cases considered Cases 2 and 3 give almost identical results and both are better than Case 1. The author tried other choices for the z_i values to substantially reduce $\overline{Z}$ with no success. Maybe the reader can find other amounts to order to obtain a solution better than both Case 2 or 3. A GPSS program for one of the simulations in this chapter is in Chap. 28.

Reference

1. J.J. Buckley, T. Feuring and Y. Hayashi: Solving Fuzzy Problems in Operations Research: Inventory Control, Soft Computing, 7(2002)121-129.

18 Oil Tanker Problem

18.1 Introduction

This model has been adapted from an example in [5]. The system is shown
in Fig. 18.1. This figure shows one cycle for an oil tanker. Let us follow
an oil tanker through one round trip starting at the unload position. This
unload place is a port on the gulf coast of the US. After unloading the ship
travels to the south Atlantic ocean with travel time T1 in Fig. 18.1. In this
chapter unit time is measured in days. T1 will be modelled by the normal
distribution. Once in the south Atlantic the ship may, or may not, experience
a storm, represented as S1. The probability of a storm is p_1, so $1 - p_1$ is the
probability of no storm. A storm causes a delay in the travel time and this
delay is modelled by the normal distribution. Once out of the south Atlantic
it goes around the southern tip of Africa and up the east coast to a port in
the middle east. No storms are to be found there and its travel time is T2
also given by the normal distribution. Arriving at port it may, or may not,
experience a waiting delay until it can load oil. This possible waiting is coded
W1 and the time delay is given by a discrete probability distribution: (1) p_{10}
is the probability of zero wait; (1) p_{11} the probability of a one day wait; and
(3) p_{12} the probability of a two day wait until docking to load oil.

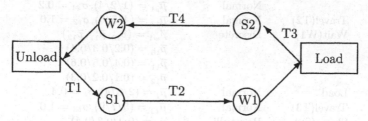

Fig. 18.1. Oil Tanker Problem

The loading time is given by the normal distribution. Once loaded the
ship travels back down the east coast of Africa (no storms) with longer travel
time T3 (normal distribution) since it is now full of oil. In the south Atlantic
the probability of a time delay (normal distribution) due to a storm (S2) is
p_2. Then it travels to the US port for unloading with travel time T4 (normal

James J. Buckley: *Simulating Fuzzy Systems*, StudFuzz **171**, 137–142 (2005)
www.springerlink.com

distribution). Once at the port it may, or may not, have to wait (W2) to unload. This possible wait is given by a discrete probability distribution: (1) p_{20} the probability of no wait; (2) p_{21} the probability of waiting one day; and (3) p_{22} the probability of a two day wait. The unloading time is modelled by the normal distribution. End of one cycle. The ships go through this round trip every day of the year.

The oil company wants to maximize yearly throughput. The company can use three types of oil tankers: (1) super tanker (ST) with capacity 30,000,000 gallons; (2) big tanker (BT) with capacity 10,000,000 gallons; and (3) regular tanker (RT) with capacity 5,000,000 gallons. So we will do three studies, one for each type of oil tanker. Some of the parameters will vary (travel time, loading/unloading time) depending on the size of the ship. Let $\Theta_{ST} = (\Theta_{BT}, \Theta_{RT})$ denote the mean number of round trips per year for the ST (BT,RT). We want to estimate the Θ values. But the company believes, due to purchase/operating costs, that one $ST \approx 2BT \approx 4RT$. These approximations are not based solely on capacities. If X is the mean gross yearly throughput we are to compare $X_{ST} = (30,000,000)\Theta_{ST}$ to $X_{BT} = (10,000,000)2\Theta_{BT}$ to $X_{RT} = (5,000,000)4\Theta_{RT}$.

Next we need to estimate all the parameters in the problem. These will be estimated from historical data and, based on Chap. 3, we will obtain fuzzy estimators. For the super tankers the results are in Table 18.1 and for the big (regular) tankers see Table 18.3 (18.5).

Table 18.1. Fuzzy Probability Distributions for the Super Tankers

Item	Distribution	Fuzzy Parameters
Unload	Normal	$\bar{\mu}_u = (2/3/4), \sigma_u = 0.4$
Travel(T1)	Normal	$\bar{\mu}_{t1} = (4/5/6), \sigma_{t1} = 0.8$
Storm(S1)	Bernoulli	$\bar{p}_1 = (0.2/0.3/0.4)$
	Normal	$\bar{\mu}_{s1} = (1/2/4), \sigma_{s1} = 0.2$
Travel(T2)	Normal	$\bar{\mu}_{t2} = (4/5/6), \sigma_{t2} = 1.0$
Wait(W1)	Discrete	$\bar{p}_{w1} = (\bar{p}_{10}, \bar{p}_{11}, \bar{p}_{12})$
		$\bar{p}_{10} = (0.2/0.3/0.4)$
		$\bar{p}_{11} = (0.4/0.5/0.6)$
		$\bar{p}_{12} = (0.1/0.2/0.3)$
Load	Normal	$\bar{\mu}_l = (2/3/4), \sigma_l = 0.4$
Travel(T3)	Normal	$\bar{\mu}_{t3} = (5/7/9), \sigma_{t3} = 1.0$
Storm(S2)	Bernoulli	$\bar{p}_2 = (0.2/0.3/0.4)$
	Normal	$\bar{\mu}_{s2} = (1/2/4), \sigma_{s2} = 0.2$
Travel(T4)	Normal	$\bar{\mu}_{t4} = (5/7/9), \sigma_{t4} = 1.0$
Wait(W2)	Discrete	$\bar{p}_{w2} = (\bar{p}_{20}, \bar{p}_{21}, \bar{p}_{22})$
		$\bar{p}_{20} = (0.1/0.2/0.3)$
		$\bar{p}_{21} = (0.3/0.4/0.5)$
		$\bar{p}_{22} = (0.3/0.4/0.5)$

We assume that there is a continuous function $F_{ST}(all\ parameters) = X_{ST}$ so that by the extension principle $\overline{X}_{ST} = F_{ST}(fuzzy\ parameters)$. Then, from Chap. 7, we may employ crisp simulation to estimate α-cuts of $\overline{X}_{ST}$. We make the same assumptions for $\overline{X}_{BT}$ and $\overline{X}_{RT}$. However, we do not know the structure of these one-step functions. Hence, we proceed to the crisp simulation. The goal is to estimate and compare the three fuzzy numbers $\overline{X}_{ST}, \overline{X}_{BT}, \overline{X}_{RT}$. These fuzzy numbers are easily found from the fuzzy numbers $\overline{\Theta}_{ST}, \overline{\Theta}_{BT}, \overline{\Theta}_{RT}$ which will be estimated from simulation.

18.2 Case 1: Super Tankers

For the super tanker the fuzzy probability distributions are shown in Table 18.1. Now we need to figure out how to choose the values of the μ_u, $\mu_{t1}, \ldots, p_{22}$ in their α-cuts to estimate the end points of the intervals for the α-cuts of $\overline{\Theta}_{ST}$. It is fairly obvious that for $max\Theta$ ($min\Theta$) we use the minimum (maximum) of μ_u, μ_{t1}, p_1, μ_{s1}, μ_{t2}, μ_l, μ_{t3}, p_2, μ_{s2}, and μ_{t4}. However, the p_{ij}, $i = 1, 2$, $j = 0, 1, 2$ in Table 18.1 are different. We may choose $p_{ij} \in \overline{p}_{ij}[\alpha]$ but we must have a discrete probability distribution $p_{i0} + p_{i1} + p_{i2} = 1$, $i = 1, 2$ [1–4]. So for $max\Theta$ ($min\Theta$) and $i = 1$ we first pick $maxp_{10}$ ($minp_{10}$) if possible, then $maxp_{11}$ ($minp_{11}$) if possible, and then $maxp_{12}$ ($minp_{12}$) if possible. The "if possible" means that you need to pick all three probabilities so that their sum is one. Similarly, choose the p_{2j}, $j = 0, 1, 2$. For example, suppose $p_{10}[0] = [0.1, 0.4]$, $p_{11}[0] = [0.3, 0.6]$ and $p_{12}[0] = [0.2, 0.5]$. Then for $max\Theta$ pick $p_{10} = 0.4$, $p_{11} = 0.4$, $p_{12} = 0.2$. We can not use $p_{11} = 0.6$ because the smallest value for p_{12} is 0.2.

Using these results we may now do the simulations. All simulations were for 50,000 round trips. For one simulation run of 50,000 cycles $CLOCK =$ total days in the simulation so $\Theta_{ST} = [50,000]/[CLOCK/365]$. Each simulation run averaged around two seconds. The results for the super tankers are in Table 18.2.

We show the graphs of $\overline{\Theta}$ in Fig. 18.2 followed by the graphs of the $\overline{X}$ in Fig. 18.3.

18.3 Case 2: Big Tankers

The big tankers load/unload and travel faster than the super tankers. This means that the values of $\overline{\mu}_u, \overline{\mu}_{t1}, \overline{\mu}_{t2}, \overline{\mu}_l, \overline{\mu}_{t3}$ and $\overline{\mu}_{t4}$ are all smaller for a BT

Table 18.2. Case 1 Simulation Alpha-Cuts of $\overline{\Theta}_{ST}$

Item	$\alpha = 0$ Cut	$\alpha = 0.5$ Cut	$\alpha = 1$ Cut
$\overline{\Theta}_{ST}$	[8.35, 15.14]	[9.50, 12.73]	10.96

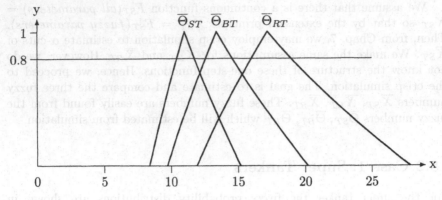

Fig. 18.2. $\overline{\Theta}$ (Trips/Year) in Cases 1–3

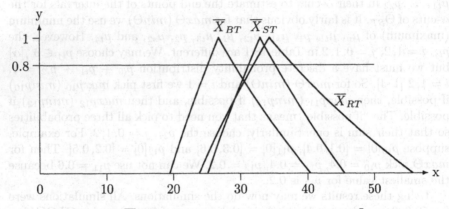

Fig. 18.3. $\overline{X}$ (Gals./Year) in Cases 1–3, Scale $= x(10)^7$ gals

than for a ST. The values of the fuzzy parameters for the big tankers are in Table 18.3. The values that do not change from Table 18.1 to Table 18.3 are those of $\overline{p}_1$, $\overline{\mu}_{s1}$, σ_{s1}, $\overline{p}_{w1}$, $\overline{p}_2$, $\overline{\mu}_{s2}$, σ_{s2} and $\overline{p}_{w2}$. Simulation results are shown in Table 18.4 with graphs in Figs. 18.2 and 18.3.

18.4 Case 3: Regular Tankers

Regular tankers load/unload and travel faster than big tankers. The values of the fuzzy parameters for regular tankers are in Table 18.5. Some values have been decreased while others are unchanged. The changes from Table 18.3 to 18.5 are similar to those from Table 18.1 to 18.3. Simulation results are given in Table 18.6 with graphs in Figs. 18.2 and 18.3.

Table 18.3. Fuzzy Probability Distributions for the Big Tankers

Item	Distribution	Fuzzy Parameters
Unload	Normal	$\overline{\mu}_u = (1/2/3), \sigma_u = 0.2$
Travel(T1)	Normal	$\overline{\mu}_{t1} = (3/4/5), \sigma_{t1} = 0.6$
Storm(S1)	Bernoulli	$\overline{p}_1 = (0.2/0.3/0.4)$
	Normal	$\overline{\mu}_{s1} = (1/2/4), \sigma_{s1} = 0.2$
Travel(T2)	Normal	$\overline{\mu}_{t2} = (3/4/5), \sigma_{t2} = 0.6$
Wait(W1)	Discrete	$\overline{p}_{w1} = (\overline{p}_{10}, \overline{p}_{11}, \overline{p}_{12})$
		$\overline{p}_{10} = (0.2/0.3/0.4)$
		$\overline{p}_{11} = (0.4/0.5/0.6)$
		$\overline{p}_{12} = (0.1/0.2/0.3)$
Load	Normal	$\overline{\mu}_l = (1/2/3), \sigma_l = 0.2$
Travel(T3)	Normal	$\overline{\mu}_{t3} = (4/6/8), \sigma_{t3} = 0.8$
Storm(S2)	Bernoulli	$\overline{p}_2 = (0.2/0.3/0.4)$
	Normal	$\overline{\mu}_{s2} = (1/2/4), \sigma_{s2} = 0.2$
Travel(T4)	Normal	$\overline{\mu}_{t4} = (4/6/8), \sigma_{t4} = 0.8$
Wait(W2)	Discrete	$\overline{p}_{w2} = (\overline{p}_{20}, \overline{p}_{21}, \overline{p}_{22})$
		$\overline{p}_{20} = (0.1/0.2/0.3)$
		$\overline{p}_{21} = (0.3/0.4/0.5)$
		$\overline{p}_{22} = (0.3/0.4/0.5)$

Table 18.4. Case 2 Simulation Alpha-Cuts of $\overline{\Theta}_{BT}$

Item	$\alpha = 0$ Cut	$\alpha = 0.5$ Cut	$\alpha = 1$ Cut
$\overline{\Theta}_{BT}$	$[9.68, 20.15]$	$[11.26, 16.10]$	13.36

18.5 Summary

Review, if needed, Sect. 2.5 of Chap. 2 on how we rank fuzzy numbers. Considering the mean number of round trips per year we see from Fig. 18.2 that regular tankers are best because $\overline{\Theta}_{ST} < \overline{\Theta}_{BT} < \overline{\Theta}_{RT}$. However, when we add in capacities and costs ($ST \approx 2BT \approx 4RT$) Fig. 18.3 shows that $\overline{X}_{BT} < \overline{X}_{ST} \approx \overline{X}_{RT}$ and ST and RT are equally best. However, $\overline{X}_{RT}$ shows more uncertainty than $\overline{X}_{ST}$ with the right end point of $\overline{X}_{RT}[0]$ much larger than the right end point of $\overline{X}_{ST}[0]$. So we might declare $\overline{X}_{RT}$ the best. These results will figure in on how the company will construct its oil tanker fleet (number of ST, BT, RT) to meet yearly oil transportation constraints between the middle east and the US. One of the GPSS programs used in this chapter is in Chap. 28.

Table 18.5. Fuzzy Probability Distributions for the Regular Tankers

Item	Distribution	Fuzzy Parameters
Unload	Normal	$\overline{\mu}_u = (0.5/1/1.5), \sigma_u = 0.1$
Travel(T1)	Normal	$\overline{\mu}_{t1} = (2/3/4), \sigma_{t1} = 0.4$
Storm(S1)	Bernoulli	$\overline{p}_1 = (0.2/0.3/0.4)$
	Normal	$\overline{\mu}_{s1} = (1/2/4), \sigma_{s1} = 0.2$
Travel(T2)	Normal	$\overline{\mu}_{t2} = (2/3/4), \sigma_{t2} = 0.4$
Wait(W1)	Discrete	$\overline{p}_{w1} = (\overline{p}_{10}, \overline{p}_{11}, \overline{p}_{12})$
		$\overline{p}_{10} = (0.2/0.3/0.4)$
		$\overline{p}_{11} = (0.4/0.5/0.6)$
		$\overline{p}_{12} = (0.1/0.2/0.3)$
Load	Normal	$\overline{\mu}_l = (0.5/1/1.5), \sigma_l = 0.1$
Travel(T3)	Normal	$\overline{\mu}_{t3} = (3/5/7), \sigma_{t3} = 0.6$
Storm(S2)	Bernoulli	$\overline{p}_2 = (0.2/0.3/0.4)$
	Normal	$\overline{\mu}_{s2} = (1/2/4), \sigma_{s2} = 0.2$
Travel(T4)	Normal	$\overline{\mu}_{t4} = (3/5/7), \sigma_{t4} = 0.6$
Wait(W2)	Discrete	$\overline{p}_{w2} = (\overline{p}_{20}, \overline{p}_{21}, \overline{p}_{22})$
		$\overline{p}_{20} = (0.1/0.2/0.3)$
		$\overline{p}_{21} = (0.3/0.4/0.5)$
		$\overline{p}_{22} = (0.3/0.4/0.5)$

Table 18.6. Case 3 Simulation Alpha-Cuts of $\overline{\Theta}_{RT}$

Item	$\alpha = 0$ Cut	$\alpha = 0.5$ Cut	$\alpha = 1$ Cut
$\overline{\Theta}_{RT}$	$[11.89, 27.84]$	$[14.09, 21.26]$	17.12

References

1. J.J. Buckley: Fuzzy Probabilities: New Approach and Applications, Physica-Verlag, Heidelberg, Germany, 2003.
2. J.J. Buckley: Fuzzy Probabilities and Fuzzy Sets for Web Planning, Springer, Heidelberg, Germany, 2004.
3. J.J. Buckley and E. Eslami: Uncertain Probabilities I: The Discrete Case, Soft Computing, 7(2003)500-505.
4. J.J. Buckley, K. Reilly and X. Zheng: Fuzzy Probabilities for Web Planning, Soft Computing, 8(2004)464-476.
5. Thomas J. Schriber: Simulation Using GPSS, John Wiley and Sons, New York, 1974.

19 Priority Queues

19.1 Introduction

This chapter will use the queuing system in Chaps. 5 and 9, but with priority orders. The system is shown in Fig. 19.1. Everything is the same as in Fig. 5.1 (9.1) except for the arrivals. Now we have three types of arrivals: (1) lowest level is called Priority #1; (2) the next lowest level is Priority #2; and (3) the highest level is labelled Priority #3. Priority determines how an item is handled in a queue. In any queue all items with priority #3 go to the front of the queue and are serviced on a first come, first served, basis. If there are no Priority #3 items in the queue, then all those with Priority #2 go to the front of the queue and are serviced in a first come, first served, manner. All items of Priority #1 are serviced in the first come, first served, mode if there are no Priority #2 or #3 items in the queue. There are different priority levels because we have different types of customers. The highest priority is given to those customers who pay in cash, so there are no billing expenses, and they place many orders. Similarly, we determine the other priority levels.

The rest of the system in Fig. 19.1 is as described in Chap. 5 (9), but we have made changes in some of the probability distributions. Arrivals are modelled using the uniform distribution and all service stations will use the normal distribution for service time. We estimate all parameters from historical data on the system, so we will obtain as in Chap. 3 fuzzy estimators. Hence, we have the fuzzy uniform (Chap. 4) for arrivals and the fuzzy normal (Chap. 4) for service time. All these fuzzy distributions are defined in Table 19.1. We also assume that all queues have unlimited capacity as assumed in Chap. 9.

The time unit in this chapter is one hour. All simulations, unless stated otherwise, will be until there are 50,000 items in the Accept-Finished box (the swamping method in Chap. 7). At the end of a simulation CLOCK will have the total hours in the study. Considering a work day to last eight hours, then CLOCK/8 gives the number of days in the simulation. Management is interested in finding R = mean time an item spends in the system and X = expected number of Accepted-Finished units per day. Let $R_i(X_i)$ be the value of $R(X)$ for Priority #i orders, $i = 1, 2, 3$. We may find the R_i from the standard output file in the simulation. But the standard output file can be made to give TX_i = the total Priority #i items that ended up in

James J. Buckley: *Simulating Fuzzy Systems*, StudFuzz **171**, 143–148 (2005)
www.springerlink.com   © Springer-Verlag Berlin Heidelberg 2005

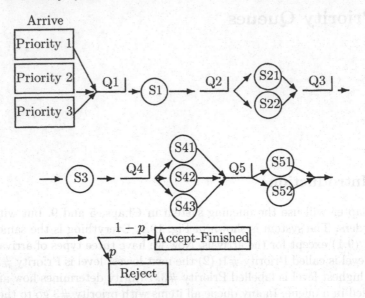

Fig. 19.1. Priority Queuing System

Table 19.1. Fuzzy Probability Distributions for Fig. 19.1

Item	Distribution	Details
Priority 1	Uniform	$\bar{a}_1 = (0.15/0.20/0.25), \bar{b}_1 = (0.35/0.40/0.45)$
Priority 2	Uniform	$\bar{a}_2 = (0.25/0.30/0.35), \bar{b}_2 = (0.45/0.50/0.55)$
Priority 3	Uniform	$\bar{a}_3 = (0.35/0.40/0.45), \bar{b}_3 = (0.55/0.60/0.65)$
Server = S1	Normal	$\bar{\mu}_1 = (0.06/0.08/0.10), \sigma_1 = 0.012$
Servers = S2	Normal	$\bar{\mu}_2 = (0.1/0.15/0.2), \sigma_2 = 0.02$
Server = S3	Normal	$\bar{\mu}_3 = (0.06/0.08/0.10).\sigma_3 = 0.012$
Servers = S4	Normal	$\bar{\mu}_4 = (0.1/0.15/0.2), \sigma_4 = 0.02$
Server = S51	Normal	$\bar{\mu}_{51} = (0.05/0.1/0.15), \sigma_{51} = 0.01$
Server = S52	Normal	$\bar{\mu}_{52} = (0.15/0.20/0.25), \sigma_{52} = 0.03$
Transfer	Bernoulli	$\bar{p} = (0.03/0.05/0.07)$

the Accept-Finished box, $i = 1, 2, 3$, for a total of 50,000 units in that box. Therefore, $X_i = TX_i/[CLOCK/8]$, $i = 1, 2, 3$.

We assume that there are continuous functions F_i (G_i) so that $R_i = F_i(all\ parameters)$ $(X_i = G_i(all\ parameters))$, $i = 1, 2, 3$. By the extension principle (Chap. 2) we obtain $\bar{R}_i = F_i(fuzzy\ parameters)$ $(\bar{X}_i = G(fuzzy\ parameters))$, $i = 1, 2, 3$. However, these functions are not available to us. Therefore, we use crisp simulation to estimate the α-cuts of the $\bar{R}_i$ and the $\bar{X}_i$ (Chap. 7).

Next we need to determine how to pick $a_1 \in \bar{a}_1[\alpha]$, $b_1 \in \bar{b}_1[\alpha], \ldots, p \in \bar{p}[\alpha]$ to approximate the end points of the intervals for $\bar{R}_k[\alpha]$ and $\bar{X}_k[\alpha]$. From

Table 19.2. Values of the Parameters for Min X_k, R_k, $k = 1, 2, 3$

Min	a_i	b_i	μ_j	p
X_k	max	max	max/min	max
R_k	max	max	min	any

Table 19.3. Values of the Parameters for Max X_k, R_k, $k = 1, 2, 3$

Max	a_i	b_i	μ_j	p
X_k	min	min	max/min	min
R_k	min	min	max	any

previous experience, plus some experimenting, the results are in Tables 19.2 and 19.3. Let us explain how we are going to use these tables. First assume that we wish to estimate the left end point of $\overline{R}_k[\alpha]$, $k = 1, 2, 3$. Table 19.2 implies that in the simulation use: (1) a_i the right end point of $\overline{a}_i[\alpha]$, $i = 1, 2, 3$; (2) b_i the right end point of $\overline{b}_i[\alpha]$, $i = 1, 2, 3$; (3) μ_j the left end point of $\overline{\mu}_j[\alpha]$, $j = 1, 2, 3, 4, 51, 52$; and (4) p any value in $\overline{p}[\alpha]$. Next consider using simulation to estimate the right end point of $\overline{X}_k[\alpha]$, $k = 1, 2, 3$. Table 19.3 says to use: (1) a_i the left end point of $\overline{a}_i[\alpha]$, $i = 1, 2, 3$; (2) b_i the left end point of $\overline{b}_i[\alpha]$, $i = 1, 2, 3$; (3) μ_j the left, or right, end point of $\overline{\mu}_j[\alpha]$, $j = 1, 2, 3, 4, 51, 52$; and (4) p the left end point of $\overline{p}[\alpha]$. We found that you get essentially the same results for $max/minX_k$ using $min\mu_j$ or $max\mu_j$, $k = 1, 2, 3$, $j = 1, 2, 3, 4, 51, 52$.

Management wants to minimize the $\overline{R}_i$ with minimizing $\overline{R}_3$ more important than minimizing the other two. Management also states that they would like to maximize the $\overline{X}_i$ with preference on maximizing $\overline{X}_3$ first, then $\overline{X}_2$ and $\overline{X}_1$ last. Let

$$\overline{Z}_r = \tau_{11}\overline{R}_1 + \tau_{12}\overline{R}_2 + \tau_{13}\overline{R}_3 , \qquad (19.1)$$

and

$$\overline{Z}_x = \tau_{21}\overline{X}_1 + \tau_{22}\overline{X}_2 + \tau_{23}\overline{X}_3 , \qquad (19.2)$$

for $\tau_{ij} > 0$, $\sum_{j=1}^{3} \tau_{ij} = 1$, $i = 1, 2$. So we want to $min\overline{Z}_r$ and $max\overline{Z}_x$. Let $M > 0$ be sufficiently large so that $min\overline{Z}_r$ is equivalent to $max[M - \overline{Z}_r]$. Finally we want to $max\overline{Z}$ where

$$\overline{Z} = \kappa_1[M - \overline{Z}_r] + \kappa_2\overline{Z}_x , \qquad (19.3)$$

for $\kappa_i > 0$ and $\kappa_1 + \kappa_2 = 1$. We decide on $\tau_{11} = 1/6$, $\tau_{12} = 2/6$, $\tau_{13} = 3/6$ and $\tau_{21} = 1/6$, $\tau_{22} = 2/6$, $\tau_{23} = 3/6$. We will use $M = 10$ and $\kappa_1 = 1/3$, $\kappa_2 = 2/3$. These are the values that we will use in this chapter to compute $\overline{Z}$ in the simulations. The emphasis was placed on Priority #3 first, then Priority #2 and lastly Priority #1 customers and on throughput $\overline{X}$ over response time $\overline{R}$.

Table 19.4. Case 1 Simulation Alpha-Cuts of $\overline{R}_k$, $\overline{X}_k$, $k = 1, 2, 3$

Item	$\alpha = 0$ Cut	$\alpha = 0.5$ Cut	$\alpha = 1$ Cut
$\overline{R}_1$	$[0.394, 0.918]$	$[0.504, 0.739]$	0.616
$\overline{R}_2$	$[0.396, 0.887]$	$[0.507, 0.737]$	0.619
$\overline{R}_3$	$[0.397, 0.873]$	$[0.508, 0.735]$	0.621
$\overline{X}_1$	$[21.33, 31.07]$	$[23.03, 27.82]$	25.22
$\overline{X}_2$	$[16.49, 22.13]$	$[17.67, 20.57]$	18.97
$\overline{X}_3$	$[13.50, 17.22]$	$[14.29, 16.18]$	15.16

19.2 Case 1: First Simulation

In this case we simulate the system in Fig. 19.1 using the data in Tables 19.1–19.3. The simulation is used to estimate the $\alpha = 0, 0.5, 1$-cuts of the fuzzy numbers $\overline{R}_k$ and $\overline{X}_k$, $k = 1, 2, 3$. The results are in Table 19.4. Using these alpha-cuts we obtain the $\alpha = 0, 0.5, 1$-cuts of $\overline{Z}_r$ (19.1), $\overline{Z}_x$ (19.2) and $\overline{Z}$ (19.3). The graphs of the $\overline{Z}$ for all cases in this chapter are in Fig. 19.2.

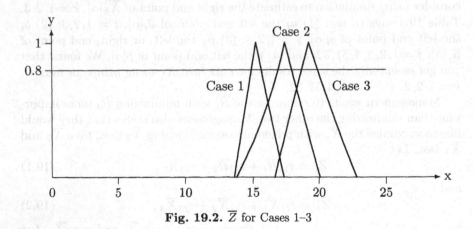

Fig. 19.2. $\overline{Z}$ for Cases 1–3

19.3 Case 2: Second Simulation

After studying the results from Case 1 management decides to consider two changes: (1) put two workers in work stations S1 and S3 because their utilization (percent time they are busy) was very high; and (2) increase the number of Priority #3 customers. The company can initiate an advertising campaign designed to attract more Priority #3 orders. The advertising company estimates, as a result of their advertising campaign, the new values of $\overline{a}_3$ and $\overline{b}_3$ to be: (1) $\overline{a}_3 = (0.20/0.25/0.30)$; and (2) $\overline{b}_3 = (0.40/0.45/0.50)$.

We now have two parallel and identical servers at S1 both with the same $\overline{\mu}_1, \sigma_1$; and there are two parallel and identical servers at S3 both with the same $\overline{\mu}_3, \sigma_3$. Using the new values of $\overline{a}_3$ and $\overline{b}_3$ we simulate the system again with results in Table 19.5 and Fig. 19.2. The * for the right end point of the $\alpha = 0$ cut means that these simulations ran until 10,000 items ended up in the Accept-Finished box (ran out of memory in the 50,000 run).

Table 19.5. Case 2 Simulation Alpha-Cuts of $\overline{R}_k, \overline{X}_k, k = 1, 2, 3$

Item	$\alpha = 0$ Cut	$\alpha = 0.5$ Cut	$\alpha = 1$ Cut
$\overline{R}_1$	$[0.391, 27.927^*]$	$[0.497, 0.725]$	0.604
$\overline{R}_2$	$[0.394, 0.921^*]$	$[0.499, 0.723]$	0.606
$\overline{R}_3$	$[0.392, 0.888^*]$	$[0.498, 0.721]$	0.604
$\overline{X}_1$	$[21.20, 29.58^*]$	$[23.15, 27.86]$	25.37
$\overline{X}_2$	$[16.54, 22.18^*]$	$[17.64, 20.42]$	18.97
$\overline{X}_3$	$[18.53, 25.85^*]$	$[20.07, 23.63]$	21.71

19.4 Case 3: Third Simulation

We see from Fig. 19.2, for our goals and how they were weighted, that Case 2 is better than Case 1. Let $\overline{Z}_i$ be the value of $\overline{Z}$ for Case i, $i = 1, 2, 3$. So far we have $\overline{Z}_1 < \overline{Z}_2$ (Sect. 2.5). Management decides on intensifying the advertising campaign to get more customers of Priority #2 and #3. However, in the production line they plan to make two changes. They will eliminate the "slower" inspector. The slower inspector will be replaced by an inspector whose service time will equal to the faster inspector, or the slower inspector will be trained to work at the same speed as the faster inspector. The inspection station will now be two parallel and identical workers with $\overline{\mu}_5 = (0.05/0.1/0.15)$, $\sigma_5 = 0.01$. Next, the utilization (percent time both workers are busy) at S2 can be very high (near 100%), so they will add another worker. Now S2 will have three parallel and identical servers with $\overline{\mu}_2$ and σ_2 given in Table 19.1.

We have not addressed the costs of all these changes. We have added a worker at S1, S2 and S3, we upgraded the service time at S52, and we have become involved in advertising campaigns. A study on how these expenses effect the unit cost of the product would be another study not undertaken in this book.

The advertising company has been asked to estimate, as a result of their new advertising campaign, values of $\overline{a}_i$ and $\overline{b}_i$, for $i = 2, 3$. They present management with the following values: (1) $\overline{a}_3 = (0.15/0.20/0.25)$ and $\overline{b}_3 = (0.35/0.40/0.45)$; and (2) $\overline{a}_2 = (0.20/0.25/0.30)$ and $\overline{b}_2 = (0.40/0.45/0.50)$.

We may now simulate the fuzzy system with results for the α-cuts in Table 19.6 and the graph of $\overline{Z}_3$ in Fig. 19.2.

Table 19.6. Case 3 Simulation Alpha-Cuts of $\overline{R}_k$, $\overline{X}_k$, $k = 1, 2, 3$

Item	$\alpha = 0$ Cut	$\alpha = 0.5$ Cut	$\alpha = 1$ Cut
$\overline{R}_1$	[0.371, 0.781]	[0.467, 0.665]	0.564
$\overline{R}_2$	[0.371, 0.779]	[0.467, 0.666]	0.565
$\overline{R}_3$	[0.371, 0.778]	[0.467, 0.664]	0.564
$\overline{X}_1$	[21.19, 31.04]	[23.15, 27.81]	25.32
$\overline{X}_2$	[18.60, 25.96]	[20.05, 23.61]	21.71
$\overline{X}_3$	[21.19, 31.06]	[23.09, 27.97]	25.26

19.5 Summary

From out three simulations we see that, from Fig. 19.2, $\overline{Z}_1 < \overline{Z}_2 < \overline{Z}_3$ and Case 3 is the best. The base of these fuzzy numbers is like a 99% confidence interval. Corporate headquarters wants this plant to accept more Priority #1 orders, from other plants around the country, because they are overloaded with this type of customer. This will produce new values for $\overline{a}_1$ and $\overline{b}_1$ for the next simulation. Also, we need to do the study on how all these changes will effect our product's unit manufacturing cost and consider a new goal of min Cost or max Profit.

20 Optimizing a Production Line

20.1 Introduction

The simple production line considered in this chapter is shown in Fig. 20.1. This problem has been adapted from an example in [1]. This situation is a little different from those in the previous chapters because items moving through the system are the workers in the production line. We begin with n workers arriving at the first station labelled A in Fig. 20.1. Our first decision variable is n and we are to find the optimal value for n. Let us call the item we are producing a "zapper". The parts, pieces, raw materials for one zapper are provided at station A and each worker uses these items to assemble one zapper. The assembly time is modelled by the normal distribution. After assembly the worker takes the zapper to the next station labelled O in the figure. At O the zapper is painted and baked and this station will be called an oven. The oven processes only one zapper at a time and the workers must queue up in front of the oven waiting to place their zapper in the oven. The service time at O is also modelled by the normal probability distribution. Our second decision variable is k which is the number of ovens to be placed at station O. Two or more ovens will be modelled as identical and parallel servers. The worker must wait at the oven until it has finished and after the zapper leaves the oven the worker places it in the finished box and returns to A to assemble another zapper. The items moving through the system are the workers.

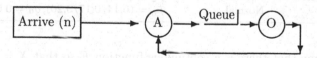

Fig. 20.1. Production Line

Time units will be in hours. We want to find n and k to maximize profit Π over a one month period. We assume the production line runs eight hours per day, five days per week, or 20 days per month. The various costs/prices involved in this product are given in Table 20.1. We see that a worker costs

James J. Buckley: *Simulating Fuzzy Systems*, StudFuzz **171**, 149–154 (2005)
www.springerlink.com

Table 20.1. Costs/Prices in the Production Line

Item	Amount
Worker	12 \$/hr.
Oven	80 \$/day
Raw Materials	2.15 \$/unit
Selling Price	20.95 \$/unit

(including fringe benefits, etc.) \$12 per hour and one oven is around \$80 per day. The parts, pieces, raw materials for one zapper is \$2.15 per unit and the selling price (and we can sell all we make at this price) is \$20.95 per unit. Let X be the number of zappers we produce per month (20 work days). Then we wish to maximize

$$\Pi = (20.95 - 2.15)X - 1920n - 20k(80) , \qquad (20.1)$$

for $n = 1, 2, 3, \ldots$ and $k = 1, 2, 3, \ldots$.

The parameters for the normal distribution will be estimated from historical data. Then, as in Chap. 3, we obtain fuzzy estimators which are given in Table 20.2. Now we have the fuzzy normal for service time and the system becomes a fuzzy system. Throughput X will be a fuzzy number making profit also a fuzzy number $\overline{\Pi}$. So we want to find the values of n and k to maximize

$$\overline{\Pi} = (18.80)\overline{X} - 1920n - 1600k . \qquad (20.2)$$

Finally, management decides it has room for at most three ovens on this production line and, it will be crowded, but they think at most ten workers can be assigned to this job. Therefore, we will $max\overline{\Pi}$ subject to $1 \leq n \leq 10$ and $k = 1, 2, 3$.

Table 20.2. Fuzzy Probability Distributions for the Production Line

Item	Distribution	Fuzzy Parameters
Assemble	Normal	$\overline{\mu}_a = (0.4/0.5/0.6), \sigma_a = 0.08$
Oven	Normal	$\overline{\mu}_0 = (0.14/0.17/0.20), \sigma_0 = 0.028$

We assume that there is a continuous function F so that $X = F(\mu_a, \mu_0, n, k)$. Then by the extension principle $\overline{X} = F(\overline{\mu}_a, \overline{\mu}_0, n, k)$. However, we do not know this function, nor can we derive it or find it in the literature. But, as explained in Chap. 7, we may use crisp simulation to estimate the α-cuts of $\overline{X}$. In this application it is fairly clear on how this is to be done. In our simulations we will use: (1) μ_a (μ_0) the left end point of $\overline{\mu}_a[\alpha]$ ($\overline{\mu}_0[\alpha]$) to estimate the right end point of $\overline{X}[\alpha]$; and (2) μ_a (μ_0) the right end point of $\overline{\mu}_a[\alpha]$ ($\overline{\mu}_0[\alpha]$) to estimate the left end point of $\overline{X}[\alpha]$.

All simulations will be for one year, or twelve 20 work day periods (the swamping method in Chap. 7). Hence, all simulations will be for 1920 hours. A simulation run took less than one second. So we divide the throughput in a simulation by twelve to get the mean output per month. Notice that this simulation has a time constraint where our other simulations usually stopped after 50,000 units were completed.

20.2 Simulations

Now we can simulate the fuzzy system and estimate the $\alpha = 0, 0.5, 1$ cuts of $\overline{X}$ and then $\overline{\Pi}$. The results for $\overline{X}$ are in Table 20.3 and for $\overline{\Pi}$ are in Table 20.4. The graphs of selected $\overline{\Pi}$ are in Figs. 20.2–20.5. Let $\overline{X}_{n,k}$ and $\overline{\Pi}_{n,k}$ be the values of these variables for a given value of n and k.

Not all the values of $\overline{X}_{n,k}$, for $n = 1, 2, \ldots, 10$ and $k = 1, 2, 3$ are shown in Table 20.3. What happens, for a fixed value of k, is that $\overline{X}_{n,k}$ increases as n increases until the utilization of the oven reaches 100% and then $\overline{X}_{n,k}$ can not substantially increase as n increases. We do not need the other values of throughput because for fixed k these values for n (where oven utilization is near 100%) will produce $max\overline{\Pi}$. Also, Table 20.4 does not have all the values of $\overline{\Pi}_{n,k}$, for $n = 1, 2, \ldots, 10$ and $k = 1, 2, 3$. This is because for fixed k, as n increases $\overline{\Pi}_{n,k}$ moves to the right and either it hits its maximum and then moves to the left, or its maximum will be at the boundary $n = 10$. What happens now is that if the utilization of the oven gets near 100% adding

Table 20.3. Simulation Alpha-Cuts of $\overline{X}_{n,k}$

n	k	$\alpha = 0$ Cut	$\alpha = 0.5$ Cut	$\alpha = 1$ Cut
1	1	[200, 297]	[218, 265]	239
2	1	[394, 578]	[428, 517]	468
3	1	[578, 836]	[626, 751]	683
4	1	[733, 1046]	[792, 944]	862
5	1	[796, 1133]	[860, 1025]	936
6	1	[801, 1145]	[866, 1034]	942
7	1	[798, 1140]	[863, 1030]	939
8	1	[801, 1145]	[866, 1034]	942
6	2	[1169, 1710]	[1269, 1532]	1388
7	2	[1340, 1947]	[1453, 1748]	1587
8	2	[1491, 2142]	[1614, 1932]	1759
9	2	[1580, 2254]	[1708, 2037]	1858
10	2	[1598, 2279]	[1727, 2060]	1878
7	3	[1389, 2048]	[1510, 1832]	1655
8	3	[1580, 2324]	[1716, 2078]	1879
9	3	[1768, 2590]	[1920, 2319]	2100
10	3	[1947, 2844]	[2112, 2551]	2311

Table 20.4. Simulation Alpha-Cuts of $\overline{\Pi}_{n,k}$ (dollars)

n	k	$\alpha = 0$ Cut	$\alpha = 0.5$ Cut	$\alpha = 1$ Cut
2	1	[1967, 5426]	[2606, 4280]	3358
3	1	[3506, 8357]	[4409, 6759]	5480
4	1	[4500, 10385]	[5610, 8467]	6926
5	1	[3765, 10100]	[4968, 8070]	6397
6	1	[1939, 8406]	[3161, 6319]	4590
7	1	[−38, 6392]	[1184, 4324]	2613
8	1	[−1901, 4566]	[−679, 2479]	750
6	2	[7257, 17428]	[9137, 14082]	11374
7	2	[8552, 19964]	[10676, 16222]	13196
8	2	[9471, 21710]	[11783, 17762]	14509
9	2	[9224, 21895]	[11630, 17816]	14450
10	2	[7642, 20445]	[10068, 16328]	12906
7	3	[7873, 20262]	[10148, 16202]	12874
8	3	[9544, 23531]	[12101, 18906]	15165
9	3	[11158, 26612]	[14016, 21517]	17400
10	3	[12604, 29467]	[15706, 23959]	19447

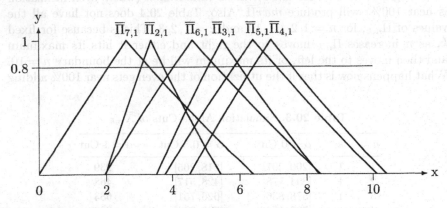

Fig. 20.2. Selected Graphs of $\overline{\Pi}_{n,1}$, One Oven, scale = $x(\$1000)$

more workers will not increase throughput, but it does increase costs, so profit decreases. So, for fixed k, we included values for $\overline{\Pi}_{n,k}$ near its maximum and sometimes omitted the rest. In this chapter $\overline{\Pi}_{n,k}$ is a discrete fuzzy set whose graph is approximated by a continuous fuzzy set (Sect. 2.7).

20.3 Summary

First we need to review Sect. 2.5 of Chap. 2 on how we rank fuzzy sets and the use of the horizontal line at 0.8 on the vertical axis. The base of the

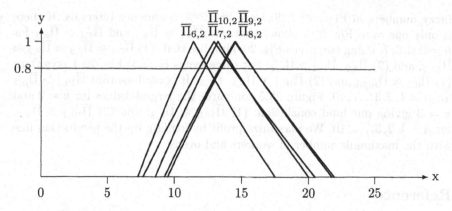

Fig. 20.3. Selected Graphs of $\overline{\overline{\Pi}}_{n,2}$, Two Ovens, scale $= x(\$1000)$

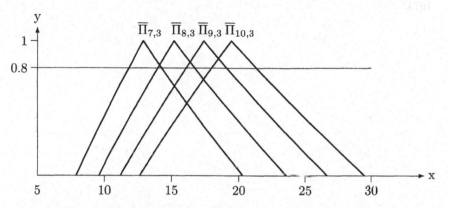

Fig. 20.4. Selected Graphs of $\overline{\overline{\Pi}}_{n,3}$, Three Ovens, scale $= x(\$1000)$

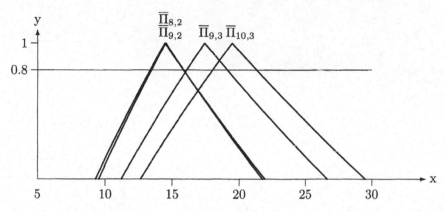

Fig. 20.5. Selected Graphs of $\overline{\overline{\Pi}}_{n,k}$, scale $= x(\$1000)$

fuzzy numbers in Figs. 20.2–20.4 are like 99% confidence intervals. If there is only one oven Fig. 20.2 shows that: (1) $\overline{\Pi}_{5,1} \approx \overline{\Pi}_{4,1}$ and $\overline{\Pi}_{4,1} > \overline{\Pi}_{n,1}$ for $n = 2, 3, 6, 7$. Using two ovens Fig. 20.3 implies that: (1) $\overline{\Pi}_{7,2} \approx \overline{\Pi}_{8,2} \approx \overline{\Pi}_{9,2} \approx \overline{\Pi}_{10,2}$; and (2) $\overline{\Pi}_{8,2}, \overline{\Pi}_{9,2} > \overline{\Pi}_{6,2}$. For three ovens ($k = 3$) Fig. 20.4 says that: (1) $\overline{\Pi}_{9,3} \approx \overline{\Pi}_{10,3}$; and (2) $\overline{\Pi}_{10,3} > \overline{\Pi}_{8,3}, \overline{\Pi}_{7,3}$. It is obvious that $\overline{\Pi}_{10,3} > \overline{\Pi}_{n,1}$ for $n = 1, 2, 3, \ldots, 10$. Figure 20.5 compares the largest values for $k = 2$ and $k = 3$ giving our final conclusion: (1) $\overline{\Pi}_{10,3} \approx \overline{\Pi}_{9,3}$; and (2) $\overline{\Pi}_{10,3} > \overline{\Pi}_{n,2}$, for $n = 1, 2, 3, \ldots, 10$. We maximize profit by loading up the production line with the maximum number of workers and ovens.

Reference

1. Thomas J. Schriber: Simulation Using GPSS, John Wiley and Sons, New York, 1974.

21 Supermarket Model

21.1 Introduction

The flow through this supermarket is shown in Fig. 21.1. This model is adapted after an example in [1]. We first describe the crisp system. Customers arrive at the store according to the exponential distribution which will generate times between arrivals. They first go to the Carts area to pick up a shopping cart. We assume that every customer gets a shopping cart. Then they proceed to the shopping aisles. The time it takes to get the shopping cart and travel to the first aisle is given by the normal distribution with mean μ_c and standard deviation σ_c.

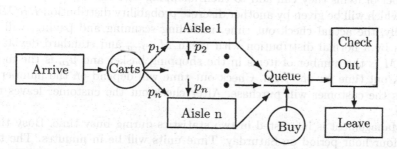

Fig. 21.1. Supermarket Model

When the shopper arrives at Aisle #1 p_1 is the probability that they decide to go down the aisle and $1 - p_1$ is the probability they skip Aisle #1 and go on to Aisle #2. At Aisle #2, independent of whether or not they went down Aisle #1, let p_2 be the probability they will travel down Aisle #2 and $1 - p_2$ is the probability they omit Aisle #2 and go on to Aisle #3. Continue this way up to Aisle #n. At Aisle #n p_n is the probability the customer will shop (go down) Aisle #n and $1 - p_n$ is the probability they skip it and move on to the check out area. The term "Aisle" is used here as a general descriptor because it can mean a real aisle in the store or: (1) the fruit and fresh vegetable area; (2) the seafood area; or (3) the meat region.

When a customer chooses to go down Aisle #i the time spent on the aisle is computed using the normal distribution and the number of items bought

James J. Buckley: *Simulating Fuzzy Systems*, StudFuzz **171**, 155–160 (2005)
www.springerlink.com

(put into the shopping cart) is given by a discrete probability distribution $PROBi$, $1 \leq i \leq m$. This normal distribution has mean μ_i and standard deviation σ_i, $1 \leq i \leq n$.

The check out area has one, two, or three check out counters open for customers to pay for their purchases. If there is only one check out counter open, then all customers must get into a single queue in order to leave the store. So now assume that there are three check out counters open. Each check out counter has its own queue and an arriving customer will always choose the shortest queue for check out. Once in a queue we assume that the customer will remain in that queue, and not jump from queue to queue, until they get to the check out counter. Let Qi be the number of customers in the i-th queue in front of the i-th check out counter at the moment a customer arrives in the check out area, $i = 1, 2, 3$. We will use the following rule for choosing the shortest queue: (1) if $Q1 \leq Q2$ is true, then choose $i = 3$ if $Q3 \leq Q1$ is true and $i = 1$ if it is false; and (2) if $Q1 \leq Q2$ is false, choose $i = 3$ if $Q3 \leq Q2$ is true and $i = 2$ if it is false. This rule is changed for only two open check out counters. We assume that all queues have unlimited capacity.

If a customer has to wait in a queue in order to check out, then they are subject to "impulse buying", like candy, a soft drink, a magazine, ... The number of items they can add to their shopping cart can be zero, one, two, etc. which will be given by another discrete probability distribution $PROB_{ib}$. Finally, the actual check out time, including scanning and paying, will be given by a normal distribution with mean $M * \mu_{co}$ and standard deviation σ_{co}. M is the number of items in the shopping basket and μ_{co} is the mean check out time for one item. Check out time will depend on the number of items the customer will purchase. After check out the customer leaves the store.

Management is interested in two statistics during busy time. Busy time is a four hour period on Saturday. Time units will be in minutes. The two statistics are: (1) the maximum number of shopping carts in use (MC); and (2) the maximum length of a queue (MQ) in front on a check out counter. They want to have a sufficient number of shopping cart on hand and they want to have enough check out counters open to keep the maximum queue length short. We are to simulate the system for 1, 2 or 3 open checkout counters and show management the results on MC and MQ for their decision (on how many check out counters to keep open).

Let us now assume that there are only six ($n = 6$) aisles in this store. It is easy to add more aisles to the model for larger stores. Table 21.1 defines all the parameters we will need for simulation. All have been estimated from historical data. However, some estimates come from national data (from corporate headquarters) and the others are store dependent. The ones from national data are the values for $Probi$, $1 \leq i \leq 6$, $Prob_{ib}$ and the p_i, $1 \leq i \leq 6$. We need to explain the notation used to define the discrete probability

Table 21.1. Fuzzy/Crisp Probability Distributions for Fig. 21.1

Item	Distribution	Details
Arrivals	Exponential	$\overline{\lambda} = (0.6/0.8/1.0)$
Carts	Normal	$\overline{\mu}_c = (2/3/4), \sigma_c = 0.4$
Aisle #1	Normal	$\overline{\mu}_1 = (4/5/6), \sigma_1 = 0.8$
Aisle #2	Normal	$\overline{\mu}_2 = (3/4/5), \sigma_2 = 0.6$
Aisle #3	Normal	$\overline{\mu}_3 = (4/6/8), \sigma_3 = 0.8$
Aisle #4	Normal	$\overline{\mu}_4 = (1/2.5/4), \sigma_4 = 0.2$
Aisle #5	Normal	$\overline{\mu}_5 = (5/7/9), \sigma_5 = 1.0$
Aisle #6	Normal	$\overline{\mu}_6 = (2/3/4), \sigma_6 = 0.4$
Check Out	Normal	$\overline{\mu}_{co} = (0.04/0.05/0.06), \sigma_{co} = 0.008$
$Prob1$	Discrete	$0.1/1, 0.2/2, 0.3/3, 0.3/4, 0.1/5$
$Prob2$	Discrete	$0.2/1, 0.5/2, 0.3/3$
$Prob3$	Discrete	$0.5/1, 0.5/2$
$Prob4$	Discrete	$0.1/2, 0.1/3, 0.2/4, 0.2/5, 0.2/6, 0.2/7$
$Prob5$	Discrete	$0.7/1, 0.3/2$
$Prob6$	Discrete	$0.5/3, 0.2/4, 0.3/5$
$Prob_{ib}$	Discrete	$0.6/0, 0.3/1, 0.1/2$
Transfer	Bernoulli	$p_1 = 0.25$
Transfer	Bernoulli	$p_2 = 0.45$
Transfer	Bernoulli	$p_3 = 0.18$
Transfer	Bernoulli	$p_4 = 0.30$
Transfer	Bernoulli	$p_5 = 0.60$
Transfer	Bernoulli	$p_6 = 0.20$

distributions $Probi$, $1 \leq i \leq 6$. Consider $Prob2$. The probability distribution is: (1) probability of one item equals 0.2; (2) the probability of buying two items is 0.5; and (3) the customer picks up three items with probability 0.3. The other parameters are computed from store data and, as in Chap. 3, we obtain fuzzy estimators for these. So we are to use the fuzzy normal and the fuzzy exponential. Notice that $\overline{\lambda}$ in Table 21.1 is a rate giving the number of arrivals per minute so $1/\overline{\lambda}$ gives the mean time between arrivals. Also notice that $\overline{\mu}_{co}$ gives the mean checkout time for one item which is approximately 3 seconds. With fuzzy probability distributions we have a fuzzy system and $\overline{MC}$ and $\overline{MQ}$ are fuzzy. Notice that $\overline{MC}$ and $\overline{MQ}$ are both discrete fuzzy sets and not fuzzy numbers. However, when we sketch the graphs of $\overline{MC}$ and $\overline{MQ}$ we will continue to use continuous curves which make them look like fuzzy numbers.

We assume that there are functions F and G so that $MC = F(all\ parameters)$ and $MQ = G(all\ parameters)$. By the extension principle $\overline{MC} = F(fuzzy\ parameters)$ and $\overline{MQ} = G(fuzzy\ parameters)$. But we do not know these functions F and G, nor can we derive them or find them in the literature. As in Chap. 7, the "easy way out" is that we will use crisp simulation to estimate alpha-cuts of $\overline{MC}$ and $\overline{MQ}$.

The next thing we need to decide on is how to choose the values of the fuzzy parameters in their alpha-cuts to estimate the α-cuts of $\overline{MC}$ and $\overline{MQ}$. Some choices were obvious and the others required some experimentations. The results are in Tables 21.2 and 21.3. From Table 21.2, to approximate the left end point of $\overline{MQ}[\alpha]$, we use: (1) λ the left end point of $\overline{\lambda}[\alpha]$; (2) μ_c the right end point of $\overline{\mu}_c[\alpha]$; (3) μ_i the right end point of $\overline{\mu}_i[\alpha]$, $1 \leq i \leq 6$; and (4) μ_{co} the left end point of $\overline{\mu}_{co}[\alpha]$. From Table 21.3, to estimate the right end point of $\overline{MC}[\alpha]$, we use: (1) λ the right end point of $\overline{\lambda}[\alpha]$; (2) μ_c the right end point of $\overline{\mu}_c[\alpha]$; (3) μ_i the right end point of $\overline{\mu}_i[\alpha]$, $1 \leq i \leq 6$; and (4) μ_{co} the right end point of $\overline{\mu}_{co}[\alpha]$.

Table 21.2. Values of the Parameters for Min MC, MQ

Min	λ	μ_c	μ_i	μ_{co}
MC	min	min	min	min
MQ	min	max	max	min

Table 21.3. Values of the Parameters for Max MC, MQ

Max	λ	μ_c	μ_i	μ_{co}
MC	max	max	max	max
MQ	max	min	min	max

All simulations will be for one year. Time units are in minutes. All simulations took less than one second. The busy time of interest was 240 minutes (four hours on Saturday) times 52 weeks equals 12,480 minutes. We simulate for one year to get into steady-state since at the start of a simulation all servers (and the store) are empty (swamping method in Chap. 7).

21.2 Case 1: First Simulation

In the first simulation we will have only one check out counter open. The results are in Table 21.4. The graphs of $\overline{MC}$ are in Fig. 21.2 and those of $\overline{MQ}$ are in Fig. 21.3.

Table 21.4. Case 1 Simulation Alpha-Cuts of $\overline{MC}$, $\overline{MQ}$

Item	$\alpha = 0$ Cut	$\alpha = 0.5$ Cut	$\alpha = 1$ Cut
$\overline{MC}$	$[19, 51]$	$[26, 45]$	35
$\overline{MQ}$	$[4, 15]$	$[6, 12]$	8

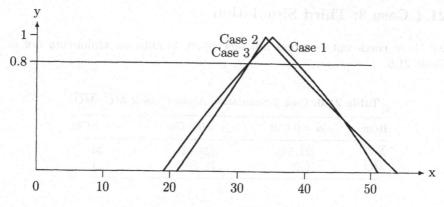

Fig. 21.2. Graphs of $\overline{MC}$, Cases 1–3

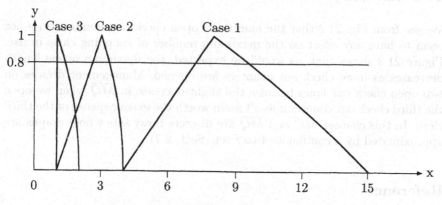

Fig. 21.3. Graphs of $\overline{MQ}$, Cases 1–3

21.3 Case 2: Second Simulation

Now we have two check out counters open. Shoppers choose the shortest queue when they arrive at the check out area. Results are in Table 21.5.

Table 21.5. Case 2 Simulation Alpha-Cuts of $\overline{MC}$, $\overline{MQ}$

Item	$\alpha = 0$ Cut	$\alpha = 0.5$ Cut	$\alpha = 1$ Cut
$\overline{MC}$	$[21, 54]$	$[28, 43]$	34
$\overline{MQ}$	$[1, 4]$	$[2, 4]$	3

21.4 Case 3: Third Simulation

All three check out counters are now open. Simulation alpha-cuts are in Table 21.6.

Table 21.6. Case 3 Simulation Alpha-Cuts of $\overline{MC}$, $\overline{MQ}$

Item	$\alpha = 0$ Cut	$\alpha = 0.5$ Cut	$\alpha = 1$ Cut
$\overline{MC}$	[21, 54]	[28, 42]	34
$\overline{MQ}$	[1, 2]	[1, 2]	1

21.5 Summary

We see from Fig. 21.2 that the number of open check out counters does not seem to have any effect on the maximum number of shopping carts in use. Figure 21.3 shows that, as would be expected, the maximum queue length decreases as more check out counters are opened. Management decides on two open check out lanes because the slight decrease in $\overline{MQ}$ when we open the third check out counter doesn't seem worth the extra expense of the third clerk. In this chapter $\overline{MC}$ and $\overline{MQ}$ are discrete fuzzy sets whose graphs are approximated by a continuous fuzzy set (Sect. 2.7).

Reference

1. Thomas J. Schriber: Simulation Using GPSS, John Wiley and Sons, New York, 1974.

22 Bank Teller Problem

22.1 Introduction

This is a classical problem in queuing theory: should there be multiple queues, one for each teller in a bank, or should we have one single queue for all the tellers? We have adopted this version of the problem from an example in [2]. The two situations are shown in Figs. 22.1 and 22.2.

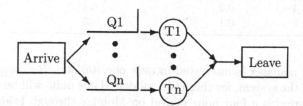

Fig. 22.1. Bank Tellers: Multiple Queues

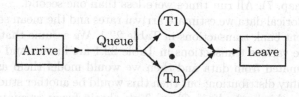

Fig. 22.2. Bank Tellers: Single Queue

First look at Fig. 22.1. Customers arrive according to the exponential distribution (which produces inter-arrival times) and they immediately go to any teller that is free (not busy). If all tellers are busy, then the customer chooses the shortest queue to wait for service. Once in a queue the customer waits there, and does not jump to other queues, in a first-come first-served basis, until he/she has completed their business. Customers require different types of transactions numbered #1 through #5 in Table 22.1. This table gives the probabilities and approximate mean service times for each of these types of transactions. For example, type #2 occurs with probability 0.3 and has

James J. Buckley: *Simulating Fuzzy Systems*, StudFuzz **171**, 161–164 (2005)
www.springerlink.com © Springer-Verlag Berlin Heidelberg 2005

approximate mean service time 2 minutes. Service times are governed by the exponential distribution. So, for transaction type #2 the service time will be computed using the exponential distribution having mean approximately 2 minutes. Customers only have one transaction, from Table 22.1, to complete and after service they leave the bank.

In Fig. 22.2 customers arrive according to the exponential distribution but they have one queue for all the tellers. The queue discipline is also first-come, first-served. The customers still have the same types of transactions, probabilities and approximate mean service times as shown in Table 22.1. Service times are computed as in the multiple queue case.

Table 22.1. Business Transactions at the Bank

Type	Probability	Mean Service Time (minutes)
1	0.1	≈1
2	0.3	≈2
3	0.3	≈3
4	0.2	≈4
5	0.1	≈5

The bank manager is interested in only one statistic, $R = $ response time, mean time in the system, for these two models. Time units will be in minutes. Our study period is a four hour period on Monday through Friday. We will simulate for one half a year, or 31,200 minutes, in order to be sure to get into steady-state since all simulations begin with the bank empty (the swamping method in Chap. 7). All run times were less than one second.

Using historical data we estimate arrival rates and the mean service times for the different bank transactions in Table 22.1. We assume that the probabilities for the various transactions in Table 22.1 are known and crisp. They could be estimated from data and then we would model them as a discrete fuzzy probability distribution; but then this would be another study which we shall not do in this book. As in Chap. 3 we obtain fuzzy estimators given in Table 22.2. Notice that in this table $\overline{\lambda}$ for arrivals is given as a rate (number of arrivals per minute), so $1/\overline{\lambda}$ is the mean time between arrivals. But the $\overline{\mu}_i$ are the mean service times in minutes, for the i-th business transaction, $1 \leq i \leq 5$. Fuzzy estimators give fuzzy distributions and fuzzy systems and then response time will be a fuzzy number $\overline{R}$. We wish to compare $\overline{R}$ for the two systems: multiple queues vs single queue.

We assume there is a continuous function F so that $R = F(all\ parameters)$ and then by the extension principle $\overline{R} = F(fuzzy\ parameters)$. However, we do not know this function, nor can we derive it or find it in the literature. So we use crisp simulation to estimate the alpha-cuts of $\overline{R}$ (Chap. 7). Actually, except for the complications of choosing the smallest queue and the different

Table 22.2. Fuzzy/Crisp Probability Distributions for the Bank

Item	Distribution	Details
Arrivals	Exponential	$\overline{\lambda}_a = (0.5/1/1.5)$ rate
Service #1	Exponential	$\overline{\mu}_1 = (0.5/1/1.5)$ time
Service #2	Exponential	$\overline{\mu}_2 = (1/2/3)$ time
Service #3	Exponential	$\overline{\mu}_3 = (2/3/4)$ time
Service #4	Exponential	$\overline{\mu}_4 = (3/4/5)$ time
Service #5	Exponential	$\overline{\mu}_5 = (4/5/6)$ time
Transactions	Discrete	$0.1/1, 0.3/2, 0.3/3, 0.2/4, 0.1/5$

transactions in Table 22.1, an F could be found in most operations research text books [1].

The last thing we need to do before simulation is to decide on how to choose the parameters in their alpha-cuts to estimate the end points of the intervals $\overline{R}[\alpha]$. In this problem this is easily solved: (1) for the left end point of $\overline{R}[\alpha]$ use λ the left end point of $\overline{\lambda}[\alpha]$ and use μ_i the left end point of $\overline{\mu}_i[\alpha]$, $1 \leq i \leq 5$; and (2) for the right end point of $\overline{R}[\alpha]$ use λ the right end point of $\overline{\lambda}[\alpha]$ and use μ_i the right end point of $\overline{\mu}_i[\alpha]$, $1 \leq i \leq 5$.

22.2 First Simulation: Multiple Queues

The simulation results are in Table 22.3 and graphs of $\overline{R}$ are in Fig. 22.3. For the rest of this chapter we assume that there are six tellers. The bank manager does not want to hire more tellers.

Table 22.3. Multiple Queues Simulation Alpha-Cuts of $\overline{R}$

Item	$\alpha = 0$ Cut	$\alpha = 0.5$ Cut	$\alpha = 1$ Cut
$\overline{R}$	$[1.976, 16.504]$	$[2.475, 4.569]$	3.180

22.3 Second Simulation: Single Queue

The simulation results are in Table 22.4 and the graph of $\overline{R}$ is in Fig. 22.3.

22.4 Summary

We see from Fig. 22.3 that there is not much difference in the time spent in the system for the two models. The base of these fuzzy numbers is like a 99%

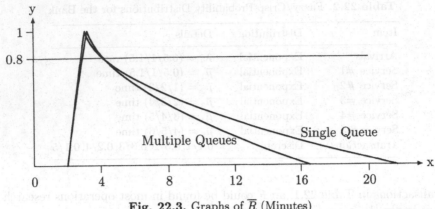

Fig. 22.3. Graphs of $\overline{R}$ (Minutes)

Table 22.4. Single Queue Simulation Alpha-Cuts of $\overline{R}$

Item	$\alpha = 0$ Cut	$\alpha = 0.5$ Cut	$\alpha = 1$ Cut
$\overline{R}$	[1.976, 21.690]	[2.456, 4.105]	3.036

confidence interval. However, multiple queue $\overline{R}$ has less uncertainty than the single queue $\overline{R}$. It looks like we almost have multiple queue $\overline{R}$ a fuzzy subset of the single queue $\overline{R}$ (Sect. 2.2.3 of Chap. 2). We would recommend the single queue method (see below).

There are situations where one model may do better (less response time) that the other model. In fact, for the multiple queue model, if we let customers change queues to the faster moving line, then its response time should decrease and the multiple queue model may be doing a little better than the single queue model.

Many banks in the US have changed to the single queue model. When asked why the change the answer was usually that it gives the customer more privacy with the teller. This bank manager sees that both models give approximately the same response time, the multiple queue method has less uncertainty but the single queue allows more privacy, so the bank manager changes to the single queue. A GPSS program for the single queue model is in Chap. 28.

References

1. H.A. Taha: Operations Research, Fifth Edition, Macmillan, New York, 1992.
2. Thomas J. Schriber: Simulation Using GPSS, John Wiley and Sons, New York, 1974.

23 A Bus Stop

23.1 Introduction

We have plenty of complaints about a certain bus stop. People can not get on the bus because when it stops it is already too crowded. We could add more buses to this bus line or maybe buy larger buses. Buying bigger buses is very expensive, so lets do a study to see if we can alleviate this problem by assigning some buses from other less crowded lines to this line. Figure 23.1 shows the set up. This example has been adapted from an example in [1].

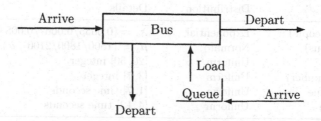

Fig. 23.1. A Bus Station

Buses arrive according to the normal distribution with mean time μ_a between arrivals approximately 30 minutes. The capacity of this bus is 50 passengers. However, the bus is usually not completely full with the number of passengers uniformly distributed between 20 and 50. When the bus arrives the first thing it does is unload passengers. The number getting off the bus is uniformly distributed between 3 and 7. The time it takes for a passenger to get off the bus is uniformly distributed between 1 and 7 seconds. In this model we will now measure time in seconds.

Once the unloading is complete we can start the loading up to capacity of 50. Loading is done on first-come first-served basis until the queue is empty or the bus is full. The time it takes for each passenger to get on the bus is uniformly distributed from 4 to 12 seconds. People arrive at this bus station according to the exponential distribution (producing inter-arrival times) at the rate λ_a of approximately 24 per hour. Those who may not get on the bus wait for the next bus.

James J. Buckley: *Simulating Fuzzy Systems*, StudFuzz **171**, 165–168 (2005)
www.springerlink.com

We want to know about the maximal queue length at the bus station. Apparently, the queues can get very long and when a bus arrives most of the people can not get on the bus. Let MQ denote the maximal queue length. We wish to estimate MQ for the morning busy time.

All simulations will run for N buses passing through this bus stop. We wish to choose N sufficiently large for steady-state conditions because when a simulation begins the queue is always empty (the swamping method Chap. 7). Let $N = 10,000$ with time in seconds. All run times were between one and two seconds.

The parameters, and discrete probability distributions, for this study are presented in Table 23.1. We assumed that the discrete uniform distributions are known and not fuzzy but the normal and exponential are fuzzy. We then have a fuzzy system and the maximum queue length will be a discrete fuzzy set $\overline{MQ}$. When we graph $\overline{MQ}$ we will use continuous curves which will make $\overline{MQ}$ look like a fuzzy number (Sect. 2.7).

Table 23.1. Fuzzy/Crisp Probability Distributions for the Bus Stop, Time in Seconds

Item	Distribution	Details
Arrivals (people)	Exponential	$\overline{\lambda}_a = (0.0055/0.0067/0.0083)$ #/sec.
Arrivals (bus)	Normal	$\overline{\mu}_a = (1500/1800/2100)$, $\sigma_a = 120$
Passengers	Uniform	$[20, 50]$ integer
Unload (number)	Uniform	$[3, 7]$ integer
Unload Time	Uniform	$[1, 7]$ time seconds
Load Time	Uniform	$[4, 12]$ time seconds

We assume that there is a function F so that $MQ = F(\text{all parameters})$ and then by the extension principle $\overline{MQ} = F(\text{fuzzy/crisp parameters})$. Since we do not know this function, nor can we derive it or find it in the literature, we will use crisp simulation to estimate its alpha-cuts (Chap. 7).

How shall we choose $\mu \in \overline{\mu}_a[\alpha]$ and $\lambda \in \overline{\lambda}_a[\alpha]$ to estimate the end points of the interval $\overline{MQ}[\alpha]$? It seems clear that choose: (1) μ (λ) the right end point of $\overline{\mu}_a[\alpha]$ ($\overline{\lambda}_a[\alpha]$) for the right end point of $\overline{MQ}[\alpha]$; and (2) μ (λ) the left end point of $\overline{\mu}_a[\alpha]$ ($\overline{\lambda}_a[\alpha]$) for the left end point of $\overline{MQ}[\alpha]$. Recall that $\overline{\lambda}_a$ is a rate so $1/\overline{\lambda}_a$ gives mean time between arrivals.

23.2 Case 1: First Simulation

Here we use the data in Table 23.1. The results are in Table 23.2 and graphs will be in Fig. 23.2. Recall that $\overline{MQ}$ is a discrete fuzzy set (Sect. 2.2.4 of Chap. 2) but we use continuous curves in its graph.

Table 23.2. Case 1 Simulation Alpha-Cuts of $\overline{MQ}$

Item	$\alpha = 0$ Cut	$\alpha = 0.5$ Cut	$\alpha = 1$ Cut
$\overline{MQ}$	$[29, 51]$	$[29, 40]$	38

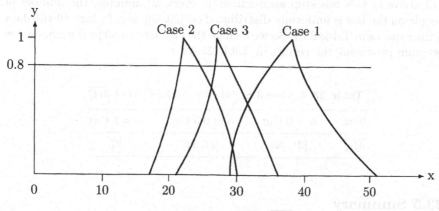

Fig. 23.2. Graphs of $\overline{MQ}$

23.3 Case 2: More Buses

From the first simulation we see that the maximum queue length is certainly too long. One possible solution is to use more buses on this bus line and we get the extra buses from other bus lines that are not as busy. Right now buses arrive at this bus stop at approximately every 30 minutes. With more buses we can make them arrive at the bus stop every 20 minutes. We will now use $\overline{\mu}_a = (900/1200/1500)$, $\sigma_a = 120$ with time in seconds. Also, the number of passengers on the bus is estimated to be uniformly distributed over $[15, 45]$. Everything else is the same as in Case 1. The simulation results are found in Table 23.3. Figure 23.2 shows a dramatic reduction in the maximum queue length using more buses.

Table 23.3. Case 2 Simulation Alpha-Cuts of $\overline{MQ}$

Item	$\alpha = 0$ Cut	$\alpha = 0.5$ Cut	$\alpha = 1$ Cut
$\overline{MQ}$	$[17, 30]$	$[21, 29]$	22

23.4 Case 3: Larger Buses

Another possible solution is to purchase larger buses. This method will be very expensive but the city council still would like to see the results for larger buses. Assume that the new buses have capacity 80 passengers. These buses will arrive at this bus stop approximately every 30 minutes, the number of people on the bus is uniformly distributed on $[20, 50]$, etc. In fact all the data is the same as in Table 23.1. So we change the capacity to 80 in the simulation program producing the results in Table 23.4.

Table 23.4. Case 3 Simulation Alpha-Cuts of $\overline{MQ}$

Item	$\alpha = 0$ Cut	$\alpha = 0.5$ Cut	$\alpha = 1$ Cut
$\overline{MQ}$	$[21, 36]$	$[27, 32]$	27

23.5 Summary

Let $\overline{MQ}_i$ be the outcome for Case $i = 1, 2, 3$. Figure 23.2 clearly shows that $\overline{MQ}_2 < \overline{MQ}_3 < \overline{MQ}_1$ (see Sect. 2.5 of Chap. 2) and we do not have to buy the bigger buses for now. We just need to transfer some buses from less busy routes to this bus route to substantially reduce the maximum queue length.

Reference

1. Thomas J. Schriber: Simulation Using GPSS, John Wiley and Sons, New York, 1974.

24 Process Failure/Spare Parts Problem

24.1 Introduction

A process in a manufacturing plant contains a very important part, which we shall now call the (spare) PART, which can fail from time to time. When the PART fails, the process in shut down, and the failed PART is sent to a repair station for repair. A failed PART can be repaired and reused. The flow through the system is shown in Fig. 24.1. This model has been adapted from an example in [1].

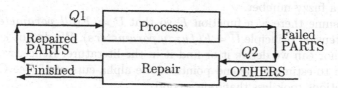

Fig. 24.1. Process Failure/Spare Parts Problem

The time to failure for the PART is given by the normal distribution with mean μ_p approximately 350 hours. The time units in this chapter will be hours. Once the PART fails, the worker assigned to this process has to remove it from the process and this time takes around 4 hours. The failed PART is taken to the queue $Q2$ in front of the repair station. The repair time for the PART is normally distributed with mean μ_r approximately 8 hours.

However, the repair facility also services other items from other processes in the plant. These items we will call the OTHERS. The arrival times for the OTHERS at the repair station is governed by the exponential distribution with mean time $1/\lambda_o$ between arrivals 9 hours. The repair time for one unit of the OTHERS is normally distributed with mean μ_o approximately 8 hours. The OTHERS have higher priority than the PARTS. This means that all units of the OTHERS go to the front of the queue $Q2$ in front of the PARTS. Items in $Q2$, with the OTHERS in front, are taken on a first-come first-served basis.

Once a PART has been repaired it goes to the queue $Q1$ of good (repaired) PARTS for the process. After the process has failed, the PART is taken out

James J. Buckley: *Simulating Fuzzy Systems*, StudFuzz **171**, 169–173 (2005)
www.springerlink.com

and sent to $Q2$, and then the worker goes to $Q1$ to get a good PART. If $Q1$ is empty he must wait for the first repaired PART. If there is a good PART in $Q1$ the worker takes it and installs it into the process. Installation time is approximately 6 hours.

Management wants to determine the utilization U of the process as a function of the number of PARTS in the system. So, we are to determine utilization U of the process for PARTS $= 1, 2, 3$, or 4. A spare part is an extra PART. If PARTS $= 1$, then spare PARTS $= 0$, and if PARTS $= 2$, then spare PARTS $= 1$, etc. All simulations will be for 10,400 hours which equals five years at 40 hours per week. The long simulation runs are to get us into steady-state since at the start of any simulation the PART is "good" and $Q2$ is empty (the swamping method in Chap. 7).

All the parameters are needed to be estimated from expert opinion (Sects. 3.2, 12.1) or from historical data. Therefore, we use fuzzy estimators for all the parameters. These are defined in Table 24.1. Notice that the two constants, time to remove (RP) and time to install (IP) a PART, become fuzzy numbers. Also, as usual, $\overline{\lambda}_o$ is given as the rate of the number of arrivals per hour. Since the constants are fuzzy numbers and we are using the fuzzy normal and the fuzzy exponential, the system becomes fuzzy and utilization $\overline{U}$ will be a fuzzy number.

We assume there is a function F so that $U = F(all\ parameters)$ and by the extension principle $\overline{U} = F(fuzzy\ parameters)$. We do not know this function, nor can we derive it or find it in the literature. Hence, we employ simulation to estimate the end points of the alpha-cuts of $\overline{U}$ (see Chap. 7). All simulations took less than one second.

The last thing to do before we can use simulation is to decide how to pick the values of the parameters in their α-cuts to estimate the end points of the intervals $\overline{U}[\alpha]$. After some experimentation, the results are in Tables 24.2 and 24.3. We found only a small effect on U from varying RP (time to remove the PART) and IP (time to install the PART). We interpret Table 24.2 to estimate the left end point of $\overline{U}[\alpha]$ as follows: (1) use the left end point of $\overline{\mu}_p[\alpha]$; and (2) use the right end point of $\overline{\lambda}_o[\alpha], \overline{\mu}_r[\alpha], \overline{\mu}_o[\alpha]$. To estimate the

Table 24.1. Fuzzy Probability Distributions for the Process Failure/Spare Parts Problem

Item	Distribution	Details
Failures	Normal	$\overline{\mu}_p = (300/350/400), \sigma_p = 60$
Arrivals (OTHERS)	Exponential	$\overline{\lambda}_o = (0.091/0.111/0.143)$
Remove PART (RP)	Constant	$RP = (3/4/5)$
Install PART (IP)	Constant	$IP = (5/6/7)$
Repair PART	Normal	$\overline{\mu}_r = (6/8/10), \sigma_r = 1.2$
Repair OTHERS	Normal	$\overline{\mu}_o = (6/8/10), \sigma_o = 1.2$

Table 24.2. Values of the Parameters for Min U

Min	λ_o	μ_p	μ_r	μ_o	RP	IP
U	max	min	max	max	max	max

Table 24.3. Values of the Parameters for Max U

Max	λ_o	μ_p	μ_r	μ_o	RP	IP
U	min	max	min	min	min	min

Table 24.4. Case 1 Simulation Alpha-Cuts of $\overline{U}$

Item	$\alpha = 0$ Cut	$\alpha = 0.5$ Cut	$\alpha = 1$ Cut
$\overline{U}$	$[0.037^*, 0.947]$	$[0.119, 0.868]$	0.727

right end point of $\overline{U}[\alpha]$ Table 24.3 implies: (1) use the right end point of $\overline{\mu}_p[\alpha]$; and (2) use the left end point of $\overline{\lambda}_o[\alpha]$, $\overline{\mu}_r[\alpha]$, $\overline{\mu}_o[\alpha]$.

24.2 Case 1: No Spare Parts

No spare parts means that the system has only one PART which starts out in the process. When it fails the worker has to wait until it is repaired and installed to start the process again. Using the fuzzy parameters in Table 24.1, and the results in Tables 24.2 and 24.3, we simulated the fuzzy system with results in Table 24.4. The value of 0.037* for the left end point of $\overline{U}[0]$ means that we ran out of memory for a simulation run of five years (10,400 hours) and cut it to 8000 hours to get simulation results for U. Graphs of $\overline{U}$ are in Fig. 24.2.

24.3 Case 2: One Spare Part

Now we have two PARTS in the system. One PART starts out in the process and the other PART is the spare PART. Otherwise, everything is the same as in Case 1. Simulation results are in Table 24.5. The value 0.074* means that the time run was only 8000 hours, otherwise for larger time runs we ran out of memory.

24.4 Case 3: Two Spare Parts

In this case there are three PARTS, one PART begins in the process with two spare PARTS in $Q1$. Simulation results are displayed in Table 24.6. The

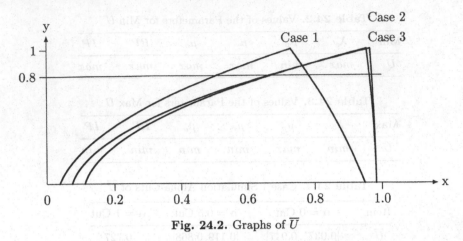

Fig. 24.2. Graphs of $\overline{U}$

Table 24.5. Case 2 Simulation Alpha-Cuts of $\overline{U}$

Item	$\alpha = 0$ Cut	$\alpha = 0.5$ Cut	$\alpha = 1$ Cut
$\overline{U}$	$[0.074^*, 0.981]$	$[0.136, 0.977]$	0.964

Table 24.6. Case 3 Simulation Alpha-Cuts of $\overline{U}$

Item	$\alpha = 0$ Cut	$\alpha = 0.5$ Cut	$\alpha = 1$ Cut
$\overline{U}$	$[0.111^*, 0.981]$	$[0.171, 0.977]$	0.953

value 0.111^* is for 7900 hours because for larger time values we ran out of memory.

24.5 Case 4: Three Spare Parts

We see from Fig. 24.2 that increasing the number of spare parts from two to three will not make an appreciable difference in the utilization of the process. So, we will omit this case.

24.6 Summary

Let $\overline{U}_i$ be the value of the fuzzy utilization of the process in Case $i = 1, 2, 3$. The base of these fuzzy numbers is like a 99% confidence interval. We see from Fig. 24.2, using $\eta = 0.8$ from Sect. 2.5 of Chap. 2, that $\overline{U}_1 \approx \overline{U}_2 \approx \overline{U}_3$.

However, if we raise η to 0.9 we obtain $\overline{U}_1 < \overline{U}_2 \approx \overline{U}_3$. Clearly, Cases 2 and 3 are the best for maximizing the utilization of the process. We would probably choose Case 2 since it is cheaper to have one less PART to buy and maintain. However, there is too much uncertainty in $\overline{U}$ for Cases 2 and 3. The left end point of $\overline{U}[0]$ is down around 10% for Cases 2 and 3. So, let us go back and consider three or more spare parts to see if this will reduce this uncertainty.

Reference

1. Thomas J. Schriber: Simulation Using GPSS, John Wiley and Sons, New York, 1974.

25 Preemptive Service

25.1 Introduction

A small city has a small municipal garage that does maintenance and repair on city owned vehicles. There are two types of vehicles: (1) TYPE 1 is the city owned cars, trucks, etc.; and (2) TYPE 2 is all the police cars. TYPE 1 vehicles get scheduled, from time to time, for maintenance and repair. All TYPE 1 vehicles scheduled for maintenance/repair on a certain day arrive before the garage opens at 8 am on that day and queue up ready for service. The number of TYPE 1 vehicles scheduled for maintenance/repair on any day is uniformly distributed between one and ten. That is, there is a random variable X with values $1, 2, \ldots, 10$ each with probability 0.1 and the value of X is the number of vehicles ready for service on any day. The city garage is small with one service bay and only one mechanic at work. Service can be performed on only one vehicle at a time. The garage is open from 8 am to 4 pm seven days per week. The 8 hours per day seven day per week requires hiring more than one mechanic, but only one mechanic is working at any time. The service time for TYPE 1 vehicles is normally distributed with mean μ_1 approximately 2 hours. Any TYPE 1 vehicle not serviced on its scheduled day waits for service on the following day. The time units in this chapter will be minutes so $\mu_1 \approx 120$.

TYPE 2 vehicles (police cars) can randomly experience a problem requiring maintence/repair any time in a 24 hour day. When this happens they have an unscheduled maintenance/repair time at the city garage. Police cars may arrive any time of the day, or night, at the garage for service. Of course, if they arrive when it is closed the car must wait until 8 am when it opens. TYPE 2 vehicles have preemptive service. What this means is: (1) if the service bay is vacant they immediately go in for service; (1) if a TYPE 1 vehicle is now being serviced it is removed from the service bay and the police car enters the service bay for service; and (3) if a TYPE 2 vehicle is now being serviced it must wait until it is finished before it can be serviced. TYPE 2 vehicles queue up for service if a TYPE 2 is now being serviced. If a TYPE 1 was removed from service for a TYPE 2, time to complete service t_c is computed, and when it returns to the service bay its service time will be t_c. Police cars arrive at the garage with inter-arrival time given by the exponential distribution with mean time $1/\lambda_2$ between arrivals 24 hours (1440

James J. Buckley: *Simulating Fuzzy Systems*, StudFuzz **171**, 175–179 (2005)
www.springerlink.com   © Springer-Verlag Berlin Heidelberg 2005

minutes). As usual, λ_2 is the rate of arrivals. Service time for TYPE 2 vehicles is normally distributed with mean μ_2 approximately 3 hours. A simple figure for this system is in Fig. 25.1. This problem has been adapted from an example in [1].

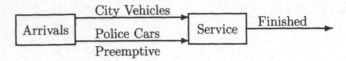

Fig. 25.1. Preemptive Service Model

The city is interested in keeping the queue lengths for both types of vehicles to a minimum. They can not have all their police cars sitting at the garage waiting for emergency maintenance/repair. Nor can they have this happen to their TYPE 1 vehicles. Let MQ_i be the maximum queue length for TYPE i vehicles, $i = 1, 2$. So, we want to find MQ_i for the present set up (described above) and consider changes (if needed) to reduce MQ_1 and/or MQ_2. There are various options to consider, but building a larger city garage is not an option.

All simulations will be for 45 days, or 64800 minutes (24 hours per day). The simulation starts at 8 am, runs for 8 hours while the shop is open, then closes for 18, to open again at 8 am the next day. Do this for 45 days. When the shop is closed vehicles arrive for service. A sufficiently long simulation run will get us into steady-state since when we start the queues will be empty (swamping method Chap. 7). All simulation runs took less than one second.

The following probabilities/parameters need to be estimated: (1) the probability of i TYPE 1 vehicles arriving for service, $i = 1, 2, \ldots, n$; (2) μ_1 and σ_1; (3) λ_2; and (4) μ_2, σ_2. We will assume that the discrete probability distribution for the arrivals of TYPE 1 vehicles is known and crisp. But we estimate the other parameters from historical data and obtain fuzzy estimators shown in Table 25.1. Recall that $\overline{\lambda}_2$ gives the number of police cars arriving at the garage per minute so $1/\overline{\lambda}_2$ is the mean time in minutes between arrivals. We now have the fuzzy normal and the fuzzy exponential implying the system is

Table 25.1. Fuzzy/Crisp Probability Distributions for the Preemptive Service Problem, Time in Minutes

Item	Distribution	Details
Arrivals (TYPE 1)	Uniform	[1, 10] integer
Service (TYPE 1)	Normal	$\overline{\mu}_1 = (100/120/140)$, $\sigma_1 = 20$
Arrivals (TYPE 2)	Exponential	$\overline{\lambda}_2 = ((1680)^{-1}/(1440)^{-1}/(1200)^{-1})$
Service (TYPE 2)	Normal	$\overline{\mu}_2 = (120/180/240)$, $\sigma_2 = 24$

Table 25.2. Values of the Parameters for Min MQ_1, MQ_2

Min	μ_1	λ_2	μ_2
MQ_1	min	min	min
MQ_2	min	min	min

Table 25.3. Values of the Parameters for Max MQ_1, MQ_2

Max	μ_1	λ_2	μ_2
MQ_1	max	max	max
MQ_2	max	max	max

fuzzy. So, the $\overline{MQ}_i$ will be discrete fuzzy sets. As usual, when we graph $\overline{MQ}_i$ we will use continuous curves connecting the base to the vertex (Sect. 2.7).

We assume that there are functions F and G so that $MQ_1 = F(all\ parameters)$ and $MQ_2 = G(all\ parameters)$. Then by the extension principle we obtain $\overline{MQ}_i$, $i = 1, 2$. But we do not know F (G), nor can we derive it or find it in the literature. Hence, we employ simulation to estimate the end points of the alpha-cuts of $\overline{MQ}_i$, $i = 1, 2$ (Chap. 7).

We need to decide on how to choose the parameters in their alpha-cuts for the simulation to estimate the end points of $\overline{MQ}_i[\alpha]$, $i = 1, 2$. The solution here is not difficult, mostly common sense, with the results shown in Tables 25.2 and 25.3.

25.2 Case 1: First Simulation

We simulate the system with parameters in Table 25.1 and using the results in Tables 25.2 and 25.3. The results are in Table 25.4 with graphs in Figs. 25.2 and 25.3.

Table 25.4. Case 1 Simulation Alpha-Cuts of $\overline{MQ}_1$, $\overline{MQ}_2$

Item	$\alpha = 0$ Cut	$\alpha = 0.5$ Cut	$\alpha = 1$ Cut
$\overline{MQ}_1$	[86, 228]	[112, 171]	141
$\overline{MQ}_2$	[3, 4]	[3, 3]	3

25.3 Case 2: Second Simulation

We see from Case 1 that the maximum queue length for TYPE 1 vehicles is absolutely too great. The values are more than the total number of TYPE

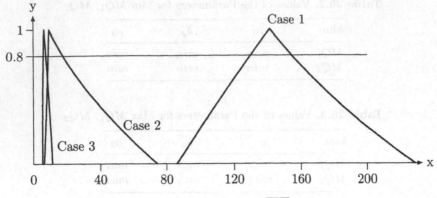

Fig. 25.2. Graphs of $\overline{MQ}_1$

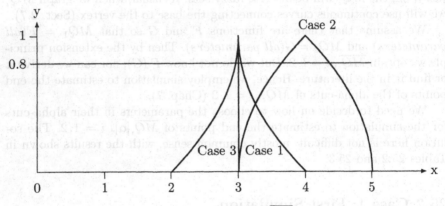

Fig. 25.3. Graphs of $\overline{MQ}_2$

1 vehicles owned by the city. Something must be done to reduce $\overline{MQ}_1$. The city decides that much of the service for its TYPE 1 vehicles can be done at local gas stations. For example, changing the oil, rotating the types, etc. can be easily and cheaply accomplished at a gas station. Hence, in the future all of this minor maintenance/repair work will be done at a local gas station. Using this information we estimate the new discrete probability distribution for the TYPE 1 vehicles arriving at the city garage each day. It is the uniform distribution on $[1, 4]$. That is, there is a random variable X with values $1, 2, 3, 4$, with each value having probability 0.25, giving the number of TYPE 1 vehicles each day. This is the only change from Case 1. Simulation results are in Table 25.5.

Table 25.5. Case 2 Simulation Alpha-Cuts of $\overline{MQ}_1$, $\overline{MQ}_2$

Item	$\alpha = 0$ Cut	$\alpha = 0.5$ Cut	$\alpha = 1$ Cut
$\overline{MQ}_1$	$[6, 74]$	$[8, 27]$	9
$\overline{MQ}_2$	$[3, 5]$	$[3, 5]$	4

25.4 Case 3: Third Simulation

We see that Case 2 produced a dramatic decrease in $\overline{MQ}_1$ but not much change in $\overline{MQ}_2$. Concentrate now on reducing $\overline{MQ}_2$. The city can sign a contract with a local garage to perform some of the more major maintenance/repair work on its police cars. This garage promises to get the job done as soon as possible. Using this information we estimate new values for $\overline{\lambda}_2$ and $\overline{\mu}_2$, σ_2. Our new fuzzy estimators are: (1) $\overline{\lambda}_2 = ((2400)^{-1}/(2160)^{-1}/(1920)^{-1})$; and (2) $\overline{\mu}_2 = (100/120/140)$, $\sigma = 20$. Everything else is the same as in Case 2. Simulation results are in Table 25.6.

Table 25.6. Case 3 Simulation Alpha-Cuts of $\overline{MQ}_1$, $\overline{MQ}_2$

Item	$\alpha = 0$ Cut	$\alpha = 0.5$ Cut	$\alpha = 1$ Cut
$\overline{MQ}_1$	$[5, 11]$	$[5, 9]$	6
$\overline{MQ}_2$	$[2, 3]$	$[3, 3]$	3

25.5 Summary

Let $\overline{MQ}_{ij}$ denote the result for TYPE i vehicle in Case j, $i = 1, 2$, $j = 1, 2, 3$. We see from Figs. 25.2 and 25.3 that $\overline{MQ}_{13} < \overline{MQ}_{12} < \overline{MQ}_{11}$ and $\overline{MQ}_{23} \approx \overline{MQ}_{21} < \overline{MQ}_{22}$ (from Sect. 2.5 in Chap. 2). We were unable to reduce $\overline{MQ}_2$ down to near one. It looks like Case 3 gives the best results. The city decides to purchase some spare police cars. A spare police car will be used when another police car is at a garage. The city officials wanted to do this anyway. A GPSS program used in this chapter is in Chap. 28.

Reference

1. Thomas J. Schriber: Simulation Using GPSS, John Wiley and Sons, New York, 1974.

26 Project Network Model

26.1 Introduction

The project network diagram is in Fig. 26.1. This problem is modelled after
an example in [2]. The project consists of various jobs that must be finished
for the project to be completed. A circle in the figure represents a node in the
project and a directed line segment joining two nodes is a job in the project.
Jobs will be written as Jij if it is a job between nodes i and j.

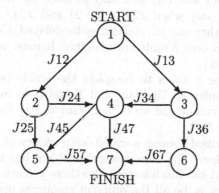

Fig. 26.1. Project Network

The figure also shows the precedence constraints on the jobs indicat-
ing which jobs must be finished before other jobs may be started. Consider
node 4. All jobs starting at node 4 can not begin until all jobs leading into
node 4 are finished. This means $J24$ and $J34$ must be completed before $J45$
and $J47$ can start.

Jobs require resources and here we will use only one resource $\mathcal{R}$ which
could be workers, money, or machines, etc. There is only so much of $\mathcal{R}$ available for the project and the number of units of $\mathcal{R}$ we have is denoted by
N. In this study N could be $5, 6, 7, \ldots, 12$. The various jobs need a minimum number of units of $\mathcal{R}$ for completion and this number is provided in
Table 26.1. Notice from Table 26.1 the maximum amount needed is 5 units
and that is why we start N at 5. Also, the time needed to finish a job is a

James J. Buckley: *Simulating Fuzzy Systems*, StudFuzz **171**, 181–186 (2005)
www.springerlink.com

Table 26.1. Resource Allocation/Job Times in the Project Network

Job	Resources Needed	Mean Time (Days)
$J12$	4 units	$\mu_{12} \approx 14$
$J13$	3 units	$\mu_{13} \approx 20$
$J24$	3 units	$\mu_{24} \approx 10$
$J25$	5 units	$\mu_{25} \approx 18$
$J34$	2 units	$\mu_{34} \approx 22$
$J36$	1 units	$\mu_{36} \approx 25$
$J45$	0 units	$\mu_{45} = 0$
$J47$	4 units	$\mu_{47} \approx 15$
$J57$	2 units	$\mu_{57} \approx 8$
$J67$	4 units	$\mu_{67} \approx 10$

random variable whose approximate mean value is also in Table 26.1. The time unit in this chapter will be days.

One job $J45$ has zero time and requires no units of $\mathcal{R}$. This is called a "dummy" job. It is put into Fig. 26.1 only to force the following job precedents: $J57$ can start only when $J25$ (node 2) and $J24$, $J34$ (node 4) are finished. There is another way job starts may be delayed. Consider a job that needs 4 units of $\mathcal{R}$ but only 3 units are available. It must wait to begin until all 4 units can be used.

Let T be the time it takes to complete the whole project and U the utilization of the resource $\mathcal{R}$. We want find T and U as functions of N and minimize T. $\mathcal{R}$ costs money and we do not want to pay for too much of the resources in reducing T.

The problem described above is a scaled down version of a much larger and very important problem. The more general problem has a project network diagram as in Fig. 26.1 but much larger. Also there is more than one resource used. Let $\mathcal{R}_i$, $1 \leq i \leq n$, be all the different resources used in the project. Each job requires a minimum of $\mathcal{R}_i$, $1 \leq i \leq n$, which could be zero for some i. We have N_i available for each $\mathcal{R}_i$, $1 \leq i \leq n$. Job times are still random according to some probability distribution. However, we may shorten a job's completion time by applying more resources. The schedule giving mean job time as a function of applied resources is known. Also the costs per unit of each resource is known. The project has a planned completion date and the company gets a bonus for each day it finishes the project before the due date. Now determine the N_i, $1 \leq i \leq n$, and the application of the resources to the jobs, to maximize profit. Most of this general problem is contained in many operations research books [1]. However, the general solution is very difficult and one usually uses simulation to explore various possible plans. Now back to our project network problem.

We will assume that job times are given by the normal distribution. The mean of the normal must be estimated from historical data and we therefore obtain a fuzzy estimator. These fuzzy estimators are given in Table 26.2. Recall that we will use a crisp estimator for the standard deviation of the normal (Chap. 8). This means that we are using the fuzzy normal for job times. The project becomes a fuzzy project with project duration $\overline{T}$ a fuzzy number and resource utilization $\overline{U}$ also a fuzzy number.

Table 26.2. Fuzzy Probability Distributions for the Project Network

Job	Distribution	Details
$J12$	Normal	$\overline{\mu}_{12} = (12/14/16)$, $\sigma_{12} = 2.4$
$J13$	Normal	$\overline{\mu}_{13} = (19/20/21)$, $\sigma_{13} = 3.8$
$J24$	Normal	$\overline{\mu}_{24} = (8/10/12)$, $\sigma_{24} = 1.6$
$J25$	Normal	$\overline{\mu}_{25} = (17/18/19)$, $\sigma_{25} = 3.2$
$J34$	Normal	$\overline{\mu}_{34} = (20/22/24)$, $\sigma_{34} = 4.0$
$J36$	Normal	$\overline{\mu}_{36} = (22/25/28)$, $\sigma_{36} = 4.4$
$J45$	Normal	$\overline{\mu}_{45} = 0$, $\sigma_{12} = NA$
$J47$	Normal	$\overline{\mu}_{47} = (13/15/17)$, $\sigma_{47} = 2.6$
$J57$	Normal	$\overline{\mu}_{57} = (7/8/9)$, $\sigma_{57} = 1.4$
$J67$	Normal	$\overline{\mu}_{67} = (6/10/14)$, $\sigma_{67} = 1.2$

We assume that there are functions F and G so that $T = F(all\ para\text{-}meters)$ and $U = G(all\ parameters)$. Then the extension principle provides $\overline{T}$ and $\overline{U}$. But we do not know F, G, nor can we derive them or find them in the literature. Hence, we use simulation to estimate the alpha-cuts of $\overline{T}$ and $\overline{U}$ (Chap. 7).

The last thing to do is to decide on how to choose $\mu_{ij} \in \overline{\mu}_{ij}[\alpha]$ to estimate the end points of $\overline{T}[\alpha]$ and $\overline{U}[\alpha]$. It seems clear that we choose μ_{ij} the left (right) end point of $\overline{\mu}_{ij}[\alpha]$ to estimate the left (right) end point of $\overline{T}[\alpha]$. It was not as clear as to what to do about $\overline{U}[\alpha]$. We found that U varied very little as we varied μ_{ij} throughout the interval $\overline{\mu}_{ij}[\alpha]$, see Table 26.3. We ended up using the same choice for the μ_{ij} as was used for $\overline{T}[\alpha]$.

26.2 Simulations

Using the values of the fuzzy estimators in Table 26.2 we simulated the fuzzy project for $N = 5, 6, \ldots, 12$. The results for the alpha-cuts of $\overline{T}$ and $\overline{U}$ are in Table 26.3. All simulations were for 10,000 runs through the project (the swamping method Chap. 7). Each simulation took less than one second of computer time. There is a slight discrepancy, due to the randomness of the simulation, for $\overline{U}$ and $N = 10$ where the left end point goes from 0.64 to 0.63 since the values should be non-decreasing. $\overline{U}$ turned out to be almost constant for fixed N so we did not produce a graph for $\overline{U}$. However, graphs for selected $\overline{T}$ are in Fig. 26.2.

Table 26.3. Simulation Alpha-Cuts of $\overline{U}$, $\overline{T}$ All Cases

Resource (N)	Item	$\alpha = 0$ Cut	$\alpha = 0.5$ Cut	$\alpha = 1$ Cut
5	$\overline{U}$	[0.72, 0.72]	[0.72, 0.72]	0.72
	$\overline{T}$	[102.01, 132.02]	[109.52, 124.52]	117.02
6	$\overline{U}$	[0.71, 0.72]	[0.71, 0.72]	0.72
	$\overline{T}$	[85.96, 110.17]	[91.90, 104.06]	97.96
7	$\overline{U}$	[0.73, 0.73]	[0.73, 0.73]	0.73
	$\overline{T}$	[71.53, 93.32]	[76.74, 87.81]	82.22
8	$\overline{U}$	[0.83, 0.84]	[0.84, 0.84]	0.84
	$\overline{T}$	[54.97, 70.68]	[58.57, 66.48]	62.42
9	$\overline{U}$	[0.69, 0.71]	[0.70, 0.71]	0.70
	$\overline{T}$	[59.06, 74.39]	[62.77, 70.46]	66.58
10	$\overline{U}$	[0.64, 0.66]	[0.63, 0.65]	0.64
	$\overline{T}$	[58.66, 72.46]	[62.10, 69.02]	65.56
11	$\overline{U}$	[0.63, 0.66]	[0.64, 0.66]	0.65
	$\overline{T}$	[52.91, 65.23]	[55.78, 61.95]	58.81
12	$\overline{U}$	[0.58, 0.61]	[0.59, 0.60]	0.60
	$\overline{T}$	[52.91, 65.23]	[55.78, 61.95]	58.81

We notice another unexpected result in Table 26.3. Look at the $\overline{T}[1]$ values as $N = 5, 6, \ldots, 12$. We see that these values decrease until $N = 8$, then increase for $N = 9$, and then decrease again. This can happen in a project network and is discussed in detail in [2].

Let $\overline{T}_i$ be the fuzzy value of project duration when $i = 5, 6, \ldots, 12$. We see from Fig. 26.2 that

$$\overline{T}_{11} \approx \overline{T}_{12} < \overline{T}_i , \qquad (26.1)$$

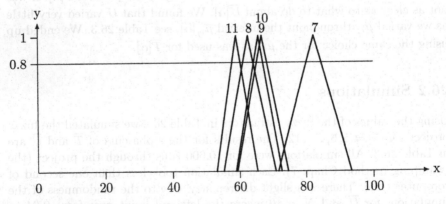

Fig. 26.2. Graphs of Selected $\overline{T}$ (Days), N Values Given

for $i = 5, 6, 7, 8, 9, 10$, even though $\overline{T}_i$, $i = 5, 6, 12$ are not shown (consult Table 26.3).

26.3 Maximize Profit

The company will receive a \$5000 bonus for each day, or fraction of a day, they finish the project in less than 70 days. However, they are to pay a penalty of \$5000 a day, or fraction of a day, that they finish in more than 70 days. The bonus/penalty is computed as \$5000$(70 - \overline{T})$. Why did the company take this contract? Consider the $\overline{\mu}_{ij}[1]$ values which are $\mu_{12} = 14$, $\mu_{13} = 20, \ldots, \mu_{67} = 10$. Using these as the job times the project duration is 57 days because the critical path (longest path through the network) goes from node #1 to node #3 to node #4 to node #7. This duration of 57 days does not consider delays due to lack of resources. But the company thinks they can surely finish in 70 days.

The other possible expense is to have unused resources (unused rental equipment). The contract will pay for used resources. Unused resources will be measured by $(1 - \overline{U})$ and the company pays for this and is not repaid for these expenses. Each unit of $\mathcal{R}$ costs \$100 per day so this cost will be computed as $100N(1 - \overline{U})\overline{T}$. So, the profit/cost of the project for the company is

$$\overline{\Pi} = 5000(70 - \overline{T}) - 100N(1 - \overline{U})\overline{T} . \tag{26.2}$$

Find N to maximize $\overline{\Pi}$.

Graphs of selected $\overline{\Pi}$ are in Fig. 26.3.

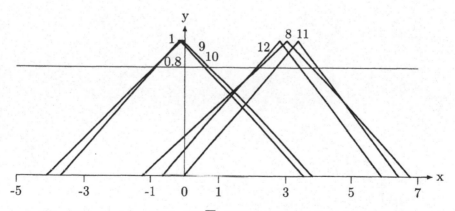

Fig. 26.3. Graphs of Selected $\overline{\Pi}$, N Values Given, Scale = x(\$10,000)

26.4 Summary

Let $\overline{\Pi}_i$ be the fuzzy profit when $i = 5, 6, \ldots, 12$. We see from Fig. 26.3 (please review Sect. 2.5 of Chap. 2) that

$$\overline{\Pi}_9 \approx \overline{\Pi}_{10} < \overline{\Pi}_8 \approx \overline{\Pi}_{11} \approx \overline{\Pi}_{12} . \tag{26.3}$$

The base of each $\overline{\Pi}_i$ is like a 99% confidence interval for profit. What shall we recommend? We recommend $N = 11$ because $\overline{\Pi}_{11}[0]$ is essentially non-negative. A GPSS program used in this chapter is in Chap. 28.

References

1. H.A. Taha: Operations Research, Fifth Edition, Macmillan, New York, 1992.
2. Thomas J. Schriber: Simulation Using GPSS, John Wiley and Sons, New York, 1974.

27 Summary and Conclusions

The first objective of this book is to explain how many systems naturally become fuzzy systems. The second objective is to show how regular (crisp) simulation can be used to estimate the alpha-cuts of the fuzzy numbers used to analyze the behavior of the fuzzy system.

Consider any system that uses probability distributions in its description. Quite often probabilities and/or parameters in the probability distributions (like the mean of the normal) have to estimated from data or expert opinion. Usually one obtains point estimates of these unknown probabilities and/or parameters for input to the system. However, it would be better to use confidence intervals. But which confidence interval should we use? We suggest in Chap. 3 to use an infinite set of confidence intervals, stacked up one on top of another, producing fuzzy number estimators. Fuzzy estimators produce fuzzy probability distributions in Chap. 4. Now the system is described by fuzzy probabilities and fuzzy probability distributions. It is a fuzzy system (Chap. 5). Things we want to compute, like response time, throughput, etc. all become fuzzy. We need to be able to compute these fuzzy sets and fuzzy numbers.

For some fuzzy systems the computation of the fuzzy descriptors becomes difficult [1,2]. We need a faster way to approximate these fuzzy numbers than multiple calculations using the extension principle. There is a method using a readily available technique called simulation. We argue in Chap. 7 that if there is a one-step function, and there usually is, that can produce the crisp form of the system descriptor, then simulation can be used to approximate the alpha-cuts of the extension principle extension of this function. Simulation may also be used to approximate some multi-step fuzzy calculations.

But to be able to use simulation to approximate the α-cuts of these fuzzy system descriptors there are certain optimization problems to solve. This is discussed in Chap. 8 and continues in Chaps . 9–26. Consider a simple example here where time between arrivals at the system is given by $N(\mu, \sigma)$, the normal distribution, and service times in the system is given by the exponential with mean λ. We want to find $X =$ throughput, or the number of transactions going through the system in an eight hour day. We estimate μ, σ and λ from data and get fuzzy estimators $\overline{\mu}$ and $\overline{\lambda}$ but we use a crisp point estimator for σ (see Chap. 8). Now find (estimate) the alpha-cuts of

James J. Buckley: *Simulating Fuzzy Systems*, StudFuzz **171**, 187–188 (2005)
www.springerlink.com   © Springer-Verlag Berlin Heidelberg 2005

fuzzy $\overline{X}$. How shall we pick $\mu \in \overline{\mu}[\alpha]$, $\lambda \in \overline{\lambda}[\alpha]$ to have simulation estimate the left (right) end point of $\overline{X}[\alpha]$? This type problem must be solved again and again in Chaps. 9–26. We are to simulate the system with specific values of μ, λ and σ to estimate the end points of alpha-cuts of the fuzzy descriptors.

We show how to use simulation to estimate the fuzzy sets/numbers needed to describe the fuzzy system in a great variety of applications in Chaps. 9–26. The applications encompass production lines, emergence rooms, inventory control and project networks.

The one part of this study that is unfinished is theoretical results for solving the simulation optimization problem (Chap. 8) discussed above. Theoretical results are needed to tell us how to choose the parameters in their alpha-cuts so that the simulation will estimate the end points of the α-cuts of the fuzzy numbers we need to analyze the fuzzy system. Sometimes we experimented with different combinations of parameter values to solve this problem. We think our results are correct but we have no theoretical justification. Eventually it would be nice to have a "look up" table to find the values of these parameters in simulations to estimate the alpha-cuts of the fuzzy numbers. That would make simulation very useful in studying fuzzy systems.

References

1. J.J. Buckley: Fuzzy Probabilities and Fuzzy Sets for Web Planning, Springer, Heidelberg, Germany, 2004.
2. J.J. Buckley, K. Reilly and X. Zheng: Fuzzy Probabilities for Web Planning, Soft Computing, 8(2004)464-476.

28 Simulation Programs

28.1 Introduction

In this chapter we present some of the GPSS programs used in Chaps. 9–26. We had to omit many programs in order to keep this chapter. less that 20 pages long. The reader may obtain other GPSS programs from the author by requesting them via email.

Most chapters run more than one GPSS simulation but here we give only one of these programs. The author does not claim that these programs are the most efficient. An expert GPSS programmer could certainly make some improvements to make it run faster. Almost all simulations ran in less than 5 seconds so such improvements are not really needed.

We do not give a tutorial on GPSS. References for GPSS were given in Chap. 6. It is not the purpose of this book to teach the reader to write programs in GPSS. These programs are presented only to show how we did the simulations. Minimal comments are given for each program written as (comment)

28.2 Chapter 9

1. SIMULATE
2. REAL &C(1),&D(1),&M(4),&V(4)
3. LET &M(1)=0.125 (Mean of Normal)
4. LET &M(2)=0.125
5. LET &M(3)=0.075
6. LET &M(4)=0.175
7. LET &V(1)=0.02 (Standard Deviation of Normal)
8. LET &V(2)=0.02
9. LET &V(3)=0.01
10. LET &V(4)=0.03
11. OUT FILEDEF 'ANS1.OUT' (File for &C(1) and &D(1))
12. STORAGE S(SHOP2),2
13. STORAGE S(SHOP4),3
14. STORAGE S(INSP),2

James J. Buckley: *Simulating Fuzzy Systems*, StudFuzz **171**, 189–204 (2005)
www.springerlink.com   © Springer-Verlag Berlin Heidelberg 2005

15. GENERATE RVEXPO(1,0.1818) (Arrivals)
16. QUEUE QUE1
17. QUEUE QUE2
18. SEIZE SHOP1
19. DEPART QUE2
20. ADVANCE 0.17,0.03 (Time in S1)
21. RELEASE SHOP1
22. TEST E S(SHOP2),0,OUT1
23. BLET &C(1)=&C(1)+1 (Counts when S2 is Idle)
24. OUT1 QUEUE QUE3
25. ENTER SHOP2
26. DEPART QUE3
27. ADVANCE RVNORM(1,&M(1),&V(1)) (Time in S2)
28. LEAVE SHOP2
29. QUEUE QUE4
30. SEIZE SHOP3
31. DEPART QUE4
32. ADVANCE 0.17,0.03 (Time in S3)
33. RELEASE SHOP3
34. TEST E S(SHOP4),0,OUT2
35. BLET &D(1)=&D(1)+1 (Counts when S4 is Idle)
36. OUT2 QUEUE QUE5
37. ENTER SHOP4
38. DEPART QUE5
39. ADVANCE RVNORM(1,&M(2),&V(2)) (Time in S4)
40. LEAVE SHOP4
41. QUEUE QUE6 (Enter the Inspection)
42. ENTER INSP
43. TEST E F(INSP1),0,OUT3 (Is Inspector #1 Free?)
44. SEIZE INSP1
45. ADVANCE RVNORM(1,&M(3),&V(3)) (Time with Inspector #1)
46. RELEASE INSP1
47. TRANSFER ,OUT4
48. OUT3 SEIZE INSP2 (Or go to Inspector #2)
49. ADVANCE RVNORM(1,&M(4),&V(4)) (Time with Inspector #2)
50. RELEASE INSP2
51. OUT4 LEAVE INSP
52. DEPART QUE6
53. DEPART QUE1
54. TRANSFER .035,,REJ (Reject with Probability p)
55. TERMINATE 1
56. REJ TERMINATE
57. START 10000

58. PUTPIC FILE=OUT,LINES=3,(&C(1),&D(1))
    SHOP2=EMPTY ****
    SHOP4=EMPTY ****
59. END

28.3 Chapter 10

1. SIMULATE
2. REAL &L(3),&M(3),&V(3),&C(1),&B(3)
3. LET &L(1)=0.2222 (Inter-arrival Times Calls A)
4. LET &L(2)=0.2857 (Calls B)
5. LET &L(3)=0.4 (Calls C)
6. LET &M(1)=0.7 (Mean of Normal for Service X)
7. LET &M(2)=2.15 (Service Y)
8. LET &M(3)=2.15 (Service Z)
9. LET &V(1)=0.08 (Standard Deviation of Normal for X)
10. LET &V(2)=0.16 (Y)
11. LET &V(3)=0.34 (Z)
12. POUT FILEDEF 'ANS2.POUT' (file POUT for &C(1),&B(3))
13. STORAGE S(CLERK1),2 (Service X)
14. STORAGE S(CLERK2),3 (Service Y)
15. STORAGE S(CLERK3),8 (Service Z)
16. GENERATE RVEXPO(2,&L(1)) (Calls A Arrive)
17. BLET PH(CALL)=1
18. TRANSFER ,OUT1
19. GENERATE RVEXPO(2,&L(2)) (Calls B Arrive)
20. BLET PH(CALL)=2
21. TRANSFER ,OUT1
22. GENERATE RVEXPO(2,&L(3)) (Calls C Arrive)
23. BLET PH(CALL)=3
24. OUT1 TEST E S(CLERK1),0,GO1
25. BLET &B(1)=&B(1)+1 (Number of Times Service X Idle)
26. GO1 TEST E S(CLERK2),0,GO2
27. BLET &B(2)=&B(2)+1 (Number of Times Service Y Idle)
28. GO2 TEST E S(CLERK3),0,GO3
29. BLET &B(3)=&B(3)+1 (Number of Times Service Z Idle)
30. GO3 TEST LE &C(1),19,REJ1 (System Full?)
31. BLET &C(1)=&C(1)+1
32. QUEUE QUE1
33. TEST E PH(CALL),3,OUT5 (Call C?)
34. TEST LE S(CLERK2),2,OUT4 (Service Y Full?)
35. TRANSFER ,OUT3
36. OUT5 TEST E PH(CALL),2,OUT6 (Call B?)

37. TEST LE S(CLERK2),2,OUT4 (Service Y Full?)
38. TRANSFER ,OUT3
39. OUT6 TEST LE S(CLERK1),1,OUT7 (Service X Full?)
40. TRANSFER ,OUT2
41. OUT7 TEST LE S(CLERK2),2,OUT4 (Service Y Full?)
42. TRANSFER ,OUT3
43. OUT2 ENTER CLERK1 (Service X)
44. BLET &C(1)=&C(1)-1
45. ADVANCE RVNORM(2,&M(1),&V(1)) (Time at X)
46. LEAVE CLERK1
47. TRANSFER ,REJ2
48. OUT3 ENTER CLERK2 (Service Y)
49. BLET &C(1)=&C(1)-1
50. ADVANCE RVNORM(2,&M(2),&V(2)) (Time at Y)
51. LEAVE CLERK2
52. TRANSFER ,REJ2
53. OUT4 ENTER CLERK3 (Service Z)
54. BLET &C(1)=&C(1)-1
55. ADVANCE RVNORM(2,&M(3),&V(3)) (Time at Z)
56. LEAVE CLERK3
57. TRANSFER ,REJ2
58. REJ2 DEPART QUE1
59. TERMINATE 1
60. REJ1 TERMINATE
61. START 10000
62. PUTPIC FILE=POUT,LINES=4,(&C(1),&B(1),&B(2),&B(3))
    NUMBER IN QUEUE ***
    CLERK1 empty *** (X)
    CLERK2 empty *** (Y)
    CLERK3 empty *** (Z)
63. END

28.4 Chapter 11

Since the system model in Chap. 11 is continued into Chap. 12 we will only
present a GPSS program used in Chap. 12.

28.5 Chapter 12

1. SIMULATE
2. REAL &M1(5),&V1(5),&M3(1),&V3(1),&L(1)
3. LET &M1(1)=1.45 (Mean of Normal)

4. LET &M1(2)=1.55
5. LET &M1(3)=1.45
6. LET &M1(4)=1.55
7. LET &M1(5)=1.45
8. LET &V1(1)=0.26 (Standard Deviation of Normal)
9. LET &V1(2)=0.28
10. LET &V1(3)=0.26
11. LET &V1(4)=0.28
12. LET &V1(5)=0.26
13. LET &M3(1)=2.0
14. LET &V3(1)=0.4
15. LET &L(1)=1.5385 (Times Between Arrivals)
16. STORAGE S(BUFFER1),50 (Queue Capacity)
17. STORAGE S(BUFFER2),50 (Queue Capacity)
18. STORAGE S(BUFFER3),50 (Queue Capacity)
19. STORAGE S(BUFFER5),50 (Queue Capacity)
20. STORAGE S(BUFFER7),50 (Queue Capacity)
21. STORAGE S(BUFFER9),50 (Queue Capacity)
22. STORAGE S(WORK),3 (M31,M32,M33)
23. GENERATE RVEXPO(1,&L(1)) (Items Arrive)
24. QUEUE QUE1
25. TEST LE S(BUFFER1),49,REJ1 (Queue Full?)
26. ENTER BUFFER1
27. ADVANCE
28. SEIZE MACH11
29. BLET PH(MACH)=1
30. LEAVE BUFFER1
31. ADVANCE RVNORM(1,&M1(1),&V1(1)) (Time in M11)
32. RELEASE MACH11
33. TRANSFER .96,,OUT1 (Go to M31,M32,M33?)
34. JUMP TEST LE S(BUFFER2),49,REJ1 (Queue Full?)
35. ENTER BUFFER2
36. ADVANCE
37. ENTER WORK
38. LEAVE BUFFER2
39. ADVANCE RVNORM(1,&M3(1),&V3(1)) (Time in M31,M32,M33)
40. LEAVE WORK
41. TRANSFER .875,,OUT11 (Reject?)
42. TRANSFER ,REJ2
43. OUT11 TEST E PH(MACH),1,OUT1 (Go Back into Production Line)
44. TEST E PH(MACH),2,OUT2 (Go Back into Production Line)
45. TEST E PH(MACH),3,OUT3 (Go Back into Production Line)
46. TEST E PH(MACH),4,OUT4 (Go Back into Production Line)
47. TRANSFER ,OUT5

48. OUT1 TEST LE S(BUFFER3),49,REJ1 (Queue Full?)
49. ENTER BUFFER3
50. ADVANCE
51. SEIZE MACH12
52. LEAVE BUFFER3
53. ADVANCE RVNORM(1,&M1(2),&V1(2)) (Time in M12)
54. RELEASE MACH12
55. TRANSFER .975,,OUT2 (Go to M31,M32,M33?)
56. TRANSFER ,JUMP
57. OUT2 TEST LE S(BUFFER5),49,REJ1 (Queue Full?)
58. ENTER BUFFER5
59. ADVANCE
60. SEIZE MACH13
61. BLET PH(MACH)=3
62. LEAVE BUFFER5
63. ADVANCE RVNORM(1,&M1(3),&V1(3)) (Time in M13)
64. RELEASE MACH13
65. TRANSFER .97,,OUT3 (Go to M31,M32,M33?)
66. TRANSFER ,JUMP
67. OUT3 TEST LE S(BUFFER7),49,REJ1 (Queue Full?)
68. ENTER BUFFER7
69. ADVANCE
70. SEIZE MACH14
71. BLET PH(MACH)=4
72. LEAVE BUFFER7
73. ADVANCE RVNORM(1,&M1(4),&V1(4)) (Time with M14)
74. RELEASE MACH14
75. TRANSFER .945,,OUT4 (Go to M31,M32,M33?)
76. TRANSFER ,JUMP
77. OUT4 TEST LE S(BUFFER9),49,REJ1 (Queue Full?)
78. ENTER BUFFER9
79. ADVANCE
80. SEIZE MACH15
81. BLET PH(MACH)=5
82. LEAVE BUFFER9
83. ADVANCE RVNORM(1,&M1(5),&V1(5)) (Time in M15)
84. RELEASE MACH15
85. TRANSFER .93,,OUT5 (Go to M31,M32,M33?)
86. TRANSFER ,JUMP
87. OUT5 DEPART QUE1
88. TERMINATE 1
89. REJ1 DEPART QUE1 (Storage)
90. TERMINATE
91. REJ2 DEPART QUE1 (Rejects)

92. TERMINATE
93. START 50000 (Simulation Run)
94. END

28.6 Chapter 13

1. SIMULATE
2. REAL &L(3),&M(6),&V(6)
3. LET &L(1)=9.0909(Time Between Arrivals Front Door B)
4. LET &L(2)=28.5714(Time Between Arrivals Front Door A)
5. LET &L(3)=28.5714(Time Between Arrivals Ambulance A)
6. LET &M(1)=9(Mean Normal)
7. LET &M(2)=11
8. LET &M(3)=5.5
9. LET &M(4)=10.5
10. LET &M(5)=27.5
11. LET &M(6)=10.5
12. LET &V(1)=1.2(Standard Deviation Normal)
13. LET &V(2)=1.6
14. LET &V(3)=0.8
15. LET &V(4)=1.8
16. LET &V(5)=4.0
17. LET &V(6)=1.8
18. STORAGE S(EVAL),2 (Size E)
19. STORAGE S(TREAT),6(Size T2)
20. STORAGE S(PW1),2(Size P1)
21. STORAGE S(SIGNIN),2(Size S)
22. GENERATE RVEXPO(1,&L(1))(Front Door B Arrives)
23. BLET PH(PAT)=1
24. TRANSFER ,OUT1
25. GENERATE RVEXPO(1,&L(2))(Front Door A Arrives)
26. BLET PH(PAT)=2
27. TRANSFER ,OUT1
28. GENERATE RVEXPO(1,&L(3))(Ambulance Arrives)
29. BLET PH(PAT)=2
30. OUT1 QUEUE QUE1
31. TEST E PH(PAT),1,OUT2
32. QUEUE QUE2
33. ENTER SIGNIN
34. DEPART QUE2
35. ADVANCE RVNORM(1,&M(1),&V(1))(Time in S)
36. LEAVE SIGNIN
37. QUEUE QUE3

38. ENTER EVAL
39. DEPART QUE3
40. ADVANCE RVNORM(1,&M(2),&V(2))(Time in E)
41. LEAVE EVAL
42. OUT2 QUEUE QUE4
43. SEIZE TRANS
44. DEPART QUE4
45. ADVANCE RVNORM(1,&M(3),&V(3))(Time in T1)
46. RELEASE TRANS
47. TEST E PH(PAT),1,OUT3
48. QUEUE QUE5
49. ENTER PW1
50. DEPART QUE5
51. ADVANCE RVNORM(1,&M(4),&V(4))(Time in P1)
52. LEAVE PW1
53. OUT3 QUEUE QUE6
54. ENTER TREAT
55. DEPART QUE6
56. ADVANCE RVNORM(1,&M(5),&V(5))(Time in T2)
57. LEAVE TREAT
58. TEST E PH(PAT),2,OUT4
59. QUEUE QUE7
60. SEIZE PW2
61. DEPART QUE7
62. ADVANCE RVNORM(1,&M(6),&V(6))(Time in P2)
63. RELEASE PW2
64. DEPART QUE1
65. TERMINATE 1
66. OUT4 DEPART QUE1
67. TERMINATE 1
68. START 50000(Run Size)
69. END

28.7 Chapter 17

1. SIMULATE
2. REAL &M(1),&V(1),&X(6),&Z(5),&W(5),&K(1),&H(1),&P(1),&D(5),
   &Y(5),&T(6), &A(1),&B(1),&Q(1),&R(1),&AA(1),&BB(1),&I,&N
3. REAL &S(6),&O(5),&BL(1),&U(1),&C(1),&CC(1)
4. LET &Z(1)=61 (Order)
5. LET &Z(2)=78 (Order)
6. LET &Z(3)=77 (Order)
7. LET &Z(4)=75 (Order)

8. LET &Z(5)=66 (Order)
9. LET &M(1)=76 (Mean of Normal)
10. LET &V(1)=10 (Standard Deviation)
11. LET &X(1)=15 (Starting Inventory)
12. LET &K(1)=101
13. LET &H(1)=22.5
14. LET &P(1)=32.5
15. LET &BL(1)=27.5 (Backorder Cost)
16. LET &A(1)=0
17. LET &B(1)=0
18. LET &C(1)=0
19. LET &N=50000
20. LET &I=1
21. OUT FILEDEF 'ANS17.OUT'(File ANS17)
22. GENERATE 10,5 (Fake Transaction)
23. BLET &D(1)=RVNORM(1,&M(1),&V(1))(Demand First Period)
24. TRANSFER .20,,OUT22 (Order Arrives?)
25. BLET &W(1)=&X(1)+&Z(1)-&D(1)(Eq 17.1, Order Arrived)
26. BLET &O(1)=0
27. TRANSFER ,OUT23
28. OUT22 BLET &W(1)=&X(1)-&D(1)(Order Late)
29. BLET &O(1)=&Z(1)
30. OUT23 TEST GE &W(1),0,OUT1 $(x_2 \geq 0?)$
31. BLET &Y(1)=&H(1)*&W(1)
32. BLET &X(2)=&W(1)
33. BLET &S(1)=0 $(s_2 = 0)$
34. TRANSFER ,OUT2
35. OUT1 BLET &Y(1)=&BL(1)*(-&W(1))
36. BLET &X(2)=0
37. BLET &S(1)=(-&W(1)) (backorder)
38. OUT2 TEST G &Z(1),0,OUT3
39. BLET &T(1)=&K(1)+10*&Z(1)+&Y(1)(Total Cost First Period)
40. TRANSFER ,OUT4
41. OUT3 BLET &T(1)=&Y(1)(Total Cost First Period)
42. ADVANCE (Comments Same All Periods)
43. OUT4 BLET &D(2)=RVNORM(1,&M(1),&V(1))
44. BLET &D(2)=&D(2)+&S(1)
45. TRANSFER .20,,OUT24
46. BLET &W(2)=&X(2)+&Z(2)+&O(1)-&D(2)
47. BLET &O(2)=0
48. TRANSFER ,OUT25
49. OUT24 BLET &W(2)=&X(2)+&O(1)-&D(2)
50. BLET &O(2)=&Z(2)
51. OUT25 TEST GE &W(2),0,OUT5

```
52.   BLET &Y(2)=&H(1)*&W(2)
53.   BLET &X(3)=&W(2)
54.   BLET &S(2)=0
55.   TRANSFER ,OUT6
56.   OUT5 BLET &Y(2)=&BL(1)*(-&W(2))
57.   BLET &X(3)=0
58.   BLET &S(2)=(-&W(2))
59.   OUT6 TEST G &Z(2),0,OUT7
60.   BLET &T(2)=&K(1)+10*&Z(2)+&Y(2)
61.   TRANSFER ,OUT8
62.   OUT7 BLET &T(2)=&Y(2)
63.   OUT8 BLET &D(3)=RVNORM(1,&M(1),&V(1))
64.   BLET &D(3)=&D(3)+&S(2)
65.   TRANSFER .20,,OUT26
66.   BLET &W(3)=&X(3)+&Z(3)+&O(2)-&D(3)
67.   BLET &O(3)=0
68.   TRANSFER ,OUT27
69.   OUT26 BLET &W(3)=&X(3)+&O(2)-&D(3)
70.   BLET &O(3)=&Z(3)
71.   OUT27 TEST GE &W(3),0,OUT9
72.   BLET &Y(3)=&H(1)*&W(3)
73.   BLET &X(4)=&W(3)
74.   BLET &S(3)=0
75.   TRANSFER ,OUT10
76.   OUT9 BLET &Y(3)=&BL(1)*(-&W(3))
77.   BLET &X(4)=0
78.   BLET &S(3)=(-&W(3))
79.   OUT10 TEST G &Z(3),0,OUT11
80.   BLET &T(3)=&K(1)+10*&Z(3)+&Y(3)
81.   TRANSFER ,OUT12
82.   OUT11 BLET &T(3)=&Y(3)
83.   OUT12 BLET &D(4)=RVNORM(1,&M(1),&V(1))
84.   BLET &D(4)=&D(4)+&S(3)
85.   TRANSFER .20,,OUT28
86.   BLET &W(4)=&X(4)+&Z(4)+&O(3)-&D(4)
87.   BLET &O(4)=0
88.   TRANSFER ,OUT29
89.   OUT28 BLET &W(4)=&X(4)+&O(3)-&D(4)
90.   BLET &O(4)=&Z(4)
91.   OUT29 TEST GE &W(4),0,OUT13
92.   BLET &Y(4)=&H(1)*&W(4)
93.   BLET &X(5)=&W(4)
94.   BLET &S(4)=0
95.   TRANSFER ,OUT14
```

96. OUT13 BLET &Y(4)=&BL(1)*(-&W(4))
97. BLET &X(5)=0
98. BLET &S(4)=(-&W(4))
99. OUT14 TEST G &Z(4),0,OUT15
100. BLET &T(4)=&K(1)+10*&Z(4)+&Y(4)
101. TRANSFER ,OUT16
102. OUT15 BLET &T(4)=&Y(4)
103. OUT16 BLET &D(5)=RVNORM(1,&M(1),&V(1))
104. BLET &D(5)=&D(5)+&S(4)
105. TRANSFER .20,,OUT30
106. BLET &W(5)=&X(5)+&Z(5)+&O(4)-&D(5)
107. BLET &O(5)=0
108. TRANSFER ,OUT31
109. OUT30 BLET &W(5)=&X(5)+&O(4)-&D(5)
110. BLET &O(5)=&Z(5)
111. OUT31 TEST GE &W(5),0,OUT17
112. BLET &Y(5)=&H(1)*&W(5)
113. BLET &X(6)=&W(5)
114. BLET &S(5)=0
115. TRANSFER ,OUT18
116. OUT17 BLET &Y(5)=&P(1)*(-&W(5))
117. BLET &X(6)=0
118. BLET §(5)=(-&W(5))
119. OUT18 TEST G &Z(5),0,OUT19
120. BLET &T(5)=&K(1)+10*&Z(5)+&Y(5)
121. TRANSFER ,OUT20
122. OUT19 BLET &T(5)=&Y(5)
123. BLET &X(6)=&X(6)+&O(5)
124. OUT20 BLET &T(6)=&T(1)+&T(2)+&T(3)+&T(4)+&T(5) (Total Cost)
125. BLET &Q(1)=&A(1)
126. BLET &A(1)=&Q(1)+&T(6)(Total Cost All Runs)
127. BLET &R(1)=&B(1)
128. BLET &B(1)=&R(1)+&X(6)(Total x_6 All Runs)
129. BLET &U(1)=&C(1)
130. BLET &C(1)=&U(1)+&S(5)(Total s_6 All Runs)
131. BLET &I=&I+1
132. TEST E &I,50000,OUT21 (counter)
133. BLET &AA(1)=&A(1)/&N (Average Cost)
134. BLET &BB(1)=&B(1)/&N (Average x_6)
135. BLET &CC(1)=&C(1)/&N (Average s_6)
136. OUT21 TERMINATE 1
137. START 50000

138. PUTPIC FILE=OUT,LINES=3,(&AA(1),&BB(1),&CC(1))
     TOTAL COST (TC)=*****
     AMT LEFT(x_6)=****
     LAST BACKORDER(s_6)=****
139. END

28.8 Chapter 18

1. SIMULATE
2. INTEGER &I
3. REAL &M(8),&V(8)
4. LET &M(1)=3.5 (Mean of Normal)
5. LET &M(2)=3.5
6. LET &M(3)=3.0
7. LET &M(4)=1.25
8. LET &M(5)=6.0
9. LET &M(6)=6.0
10. Let &M(7)=3.0
11. Let &M(8)=1.25
12. LET &V(1)=0.4 (Standard Deviation of Normal)
13. LET &V(2)=0.4
14. LET &V(3)=0.2
15. LET &V(4)=0.1
16. LET &V(5)=0.6
17. LET &V(6)=0.6
18. LET &V(7)=0.2
19. LET &V(8)=0.1
20. PROB1 FUNCTION RN(1),D3 (Probability for W1)
21. 0.25,0/0.75,1/1.0,2
22. PROB2 FUNCTION RN(1),D3 (Probability for W2)
23. 0.15,0/0.55,1/1.0,2
24. GENERATE ,,,1 (Tanker)
25. OUT5 SEIZE TRAVEL1
26. ADVANCE RVNORM(1,&M(1),&V(1)) (Time in T1)
27. RELEASE TRAVEL1
28. TRANSFER .35,,OUT1 (Storm?)
29. SEIZE TRAVEL2 (No Storm)
30. ADVANCE RVNORM(1,&M(2),&V(2))(Time in T2)
31. RELEASE TRAVEL2
32. TRANSFER ,OUT2
33. OUT1 SEIZE STORM1 (Storm)
34. ADVANCE RVNORM(1,&M(3),&V(3))(Storm Delay)
35. RELEASE STORM1

36. SEIZE TRAVEL2
37. ADVANCE RVNORM(1,&M(2),&V(2)) (Time in T2)
38. RELEASE TRAVEL2
39. OUT2 SEIZE WAIT1 (W1)
40. ADVANCE FN(PROB1)
41. RELEASE WAIT1
42. SEIZE LOAD
43. ADVANCE RVNORM(1,&M(4),&V(4))(Time to Load)
44. RELEASE LOAD
45. SEIZE TRAVEL3
46. ADVANCE RVNORM(1,&M(5),&V(5)) (Time in T3)
47. RELEASE TRAVEL3
48. TRANSFER .35,,OUT3 (Storm?)
49. SEIZE TRAVEL4 (No Storm)
50. ADVANCE RVNORM(1,&M(6),&V(6)) (Time in T4)
51. RELEASE TRAVEL4
52. TRANSFER ,OUT4
53. OUT3 SEIZE STORM2 (Storm)
54. ADVANCE RVNORM(1,&M(7),&V(7)) (Storm Delay)
55. RELEASE STORM2
56. SEIZE TRAVEL4
57. ADVANCE RVNORM(1,&M(6),&V(6)) (Time in T4)
58. RELEASE TRAVEL4
59. OUT4 SEIZE WAIT2 (W2)
60. ADVANCE FN(PROB2)
61. RELEASE WAIT2
62. SEIZE UNLOAD
63. ADVANCE RVNORM(1,&M(8),&V(8)) (Time to Unload)
64. RELEASE UNLOAD
65. BLET &I=&I+1
66. TEST GE &I,50000,OUT5 (Counter)
67. TERMINATE 1
68. START 1
69. END

28.9 Chapter 22

1. SIMULATE
2. REAL &AL(1),&L(5),&B(1)
3. LET &AL(1)=0.8 (Arrival Times)
4. LET &L(1)=1.25 (Service #1 Time)
5. LET &L(2)=2.5 (Service #2 Time)
6. LET &L(3)=3.5 (Service #3 Time)

7. LET &L(4)=4.5 (Service #4 Time)
8. LET &L(5)=5.5 (Service #5 Time)
9. STORAGE S(TELLER),6 (Six Tellers)
10. PROB1 FUNCTION RN(1),D5 (Probabilities in Table 22.1)
11. 0.1,1/0.4,2/0.7,3/0.9,4/1.0,5
12. GENERATE RVEXPO(1,&AL(1)) (Arrivals)
13. BLET PH(TIME)=FN(PROB1) (Which Type Transaction?)
14. QUEUE QUE1 (Enter Single Queue)
15. ENTER TELLER
16. BLET &B(1)=&L(PH(TIME))
17. ADVANCE RVEXPO(1,&B(1)) (Time with Teller)
18. LEAVE TELLER
19. DEPART QUE1 (Leave)
20. TERMINATE
21. GENERATE 31200
22. TERMINATE 1
23. START 1
24. END

28.10 Chapter 25

1. SIMULATE
2. REAL &M(2),&V(2),&L(1)
3. LET &L(1)=1440 (Arrival Times TYPE 2)
4. LET &M(1)=120 (Mean of Normal)
5. LET &M(2)=180 (Mean of Normal)
6. LET &V(1)=20 (Standard Deviation Normal)
7. LET &V(2)=24 (Standard Deviation Normal)
8. PROB1 FUNCTION RN(1),D10
9. 0.1,1/0.2,2/0.3,3/0.4,4/0.5,5/0.6,6/
10. 0.7,7/0.8,8/0.9,9/1.0,10 (Number of TYPE 1 Arrivals)
11. GENERATE 1440,,1,,2 (Type 2 Arrives)
12. SPLIT FN(PROB1),OUT1 (Bring in All TYPE 1)
13. TRANSFER ,OUT2
14. OUT1 QUEUE QUE1
15. SEIZE SERVICE
16. ADVANCE RVNORM(1,&M(1),&V(1)) (Service Time TYPE 1)
17. RELEASE SERVICE
18. DEPART QUE1
19. OUT2 TERMINATE
20. GENERATE RVEXPO(1,&L(1)) (Type 2 Arrives)
21. PRIORITY 2 (Top Priority)
22. QUEUE QUE2

23. PREEMPT SERVICE (Take Bay from a TYPE 1)
24. ADVANCE RVNORM(1,&M(2),&V(2)) (Service Time TYPE 2)
25. RETURN SERVICE
26. DEPART QUE2
27. TERMINATE
28. GENERATE 1440,,481,,3 (Close After 8 hours)
29. PREEMPT SERVICE,PR (Stop Service)
30. ADVANCE 960 (Let 18 Hours Pass)
31. RETURN SERVICE
32. TERMINATE
33. GENERATE 64800(Time Counter)
34. TERMINATE 1
35. START 1
36. END

28.11 Chapter 26

1. SIMULATE
2. REAL &M(9),&V(9)
3. LET &M(1)=13 (Mean J12)
4. LET &M(2)=19.5 (Mean J13)
5. LET &M(3)=9 (Mean J24)
6. LET &M(4)=17.5 (Mean J25)
7. LET &M(5)=21 (Mean J34)
8. LET &M(6)=23.5 (Mean J36)
9. LET &M(7)=14 (Mean J47)
10. LET &M(8)=7.5 (Mean J57)
11. LET &M(9)=8 (Mean J67)
12. LET &V(1)=2.4 (Standard Deviation Normal)
13. LET &V(2)=3.8
14. LET &V(3)=1.6
15. LET &V(4)=3.2
16. LET &V(5)=4.0
17. LET &V(6)=4.4
18. LET &V(7)=2.6
19. LET &V(8)=1.4
20. LET &V(9)=1.2
21. STORAGE S(MEN),12 (12 Units Available)
22. INITIAL LC(NEXT)
23. GENERATE
24. GATE LC NEXT
25. QUEUE QUE1
26. LOGIC S NEXT

27. NODE1 SPLIT 1,SUB13 SUB12
28. ENTER MEN,4 (J12 Needs 4 Units)
29. ADVANCE RVNORM(1,&M(1),&V(1)) (J12 Time)
30. LEAVE MEN,4
31. NODE2 SPLIT 1,SUB24 SUB25
32. ENTER MEN,5 (J25 Needs 5 Units)
33. ADVANCE RVNORM(1,&M(4),&V(4)) (J25 Time)
34. LEAVE MEN,5
35. NODE5 ASSEMBLE 2 SUB57
36. ENTER MEN,2 (J57 Needs 2 Units)
37. ADVANCE RVNORM(1,&M(8),&V(8)) (J57 Time)
38. LEAVE MEN,2
39. TRANSFER ,NODE7
40. SUB24 ENTER MEN,3 (J24 Needs 3 Units)
41. ADVANCE RVNORM(1,&M(3),&V(3)) (J24 Time)
42. LEAVE MEN,3
43. NODE4 ASSEMBLE 2
44. SPLIT 1,NODE5
45. SUB47 ENTER MEN,4 (J47 Needs 4 Units)
46. ADVANCE RVNORM(1,&M(7),&V(7)) (J47 Time)
47. LEAVE MEN,4
48. NODE7 ASSEMBLE 3
49. DEPART QUE1
50. LOGIC C NEXT
51. TERMINATE 1
52. SUB13 ENTER MEN,3 (J13 Needs 3 Units)
53. ADVANCE RVNORM(1,&M(2),&V(2)) (J13 Time)
54. LEAVE MEN,3
55. NODE3 SPLIT 1,SUB34 SUB36
56. ENTER MEN (J36 Needs 1 Unit)
57. ADVANCE RVNORM(1,&M(6),&V(6)) (J36 Time)
58. LEAVE MEN
59. NODE6 ENTER MEN,4 SUB67 (J67 Needs 4 units)
60. ADVANCE RVNORM(1,&M(9),&V(9)) (J67 Time)
61. LEAVE MEN,4
62. TRANSFER ,NODE7
63. SUB34 ENTER MEN,2 (J34 Needs 2 Units)
64. ADVANCE RVNORM(1,&M(5),&V(5)) (J34 Time)
65. LEAVE MEN,2
66. TRANSFER ,NODE4
67. START 10000 (Run Total)
68. END

A lot of statements are to enforce the logic in the project network, see Fig. 26.1.

Index

Printing: Strauss GmbH, Mörlenbach
Binding: Schäffer, Grünstadt